T0277474

CREATING MEANINGFUL IMPACT

Julie Bayley's book, Creating Meaningful Impact, is an enlightening romp through the excitement, the pressures, the demands of doing impact well, both in terms of institutional success and in terms of a researcher's personal and professional development. As book blurbs often suggest, the book is a rollercoaster, but one very much aimed at the fainthearted, who stand to learn a lot from Julie's immense expertise, warmth, wit and superlative use of imagery. So, if you are tickled by the idea of becoming a more mindfully impactful researcher, swipe right on 'Impact Tinder' and read this book!

–**Professor Ele Belfiore**, Professor in Cultural Policy & Director of the Interdisciplinary Centre for Social Inclusion and Cultural Diversity, University of Aberdeen, UK

Julie Bayley never fails to achieve impact on impact. If you are already on your journey to impact literacy this book will help you grow roots into impact healthy practices. And if you are just starting out, this book will help you sow the seeds that will grow into those roots to sustain your career of research with an impact on society. 'Creating meaningful impact' isn't just the title, it is the goal that Julie achieves in this important book.

–**Dr David Phipps**, Assistant VP Research Strategy & Impact, York University, Canada, and Director of Research Impact Canada

There are many books available to advise researcher how to 'do' impact but none as accessible as this. The sheer joy and enthusiasm that Julie brings to the field shines through every word which, along with insights from other researchers and partners in the field, ensures that every reader will emerge from this book enlightened, and excited about the prospect of pursuing their own 'societal impact'.

–**Dr Gemma Derrick**, Associate Professor, Research Policy & Culture, University of Bristol, UK

CREATING MEANINGFUL IMPACT: THE ESSENTIAL GUIDE TO DEVELOPING AN IMPACT-LITERATE MINDSET

BY

JULIE BAYLEY
University of Lincoln, UK

United Kingdom – North America – Japan – India
Malaysia – China

Emerald Publishing Limited
Howard House, Wagon Lane, Bingley BD16 1WA, UK

First edition 2023

Copyright © 2023 Julie Bayley.
Published under exclusive license by Emerald Publishing Limited.

Reprints and permissions service
Contact: permissions@emeraldinsight.com

No part of this book may be reproduced, stored in a retrieval system,
transmitted in any form or by any means electronic, mechanical,
photocopying, recording or otherwise without either the prior written
permission of the publisher or a licence permitting restricted copying
issued in the UK by The Copyright Licensing Agency and in the USA
by The Copyright Clearance Center. No responsibility is accepted
for the accuracy of information contained in the text, illustrations
or advertisements. The opinions expressed in these chapters are not
necessarily those of the Author or the publisher.

British Library Cataloguing in Publication Data
A catalogue record for this book is available from the British Library

ISBN: 978-1-80455-192-9 (Print)
ISBN: 978-1-80455-189-9 (Online)
ISBN: 978-1-80455-191-2 (Epub)

ISOQAR certified
Management System,
awarded to Emerald
for adherence to
Environmental
standard
ISO 14001:2004.

ISOQAR
REGISTERED
Certificate Number 1985
ISO 14001

INVESTOR IN PEOPLE

CONTENTS

List of Figures and Tables   xi

About the Author   xiii

Acknowledgements   xv

Contributors   xvii

Introduction   1

    Structure of the Book   3

Part 1: Impact, Impact Literacy and Values

Chapter 1: What Is Research Impact?   9

    What Impact Is   9

    What Impact Isn't   13

    Types of Impact   14

    Proving Impact   16

    Why Do We Do Impact?   17

        Impact in Funding   18

        Impact in Assessment   20

        Impact in Missions   22

        Personal Motivation for Impact   24

    The Wonderful World of Impact Terminology   25

        Things That Sound Like Research Impact But Aren't   30

    Dimensions of Impact   34

        Significance: How Important Is it to the Outside World?   34

        Reach: How Far or Deep Is the Effect?   35

Contribution and Attribution: How Much of the Change is
Down to the Research?   37

Distance and Time: Where and When Does the
Impact Happen?   38

Linearity and Dependencies: How Sequenced Does it
Have to Be?   39

Disciplinary Differences? *Not as Such*   41

   Fundamental or 'Discovery' Research   41

   Philosophical Research   42

   Participatory or Engaged Research   42

   Research Which Aims to Develop a Useful 'Thing'   42

   Research in Contested, Sensitive, Taboo or
   Secret Areas   43

   Commissioned Research   44

   Research to Curate, Preserve or Order Knowledge   45

What Counts as 'Better' Impact? *(If You Need to Pick)*   45

Summary   47

What Can You Do?   48

Chapter 2: Impact Literacy   51

   What Is Impact Literacy?   53

   Evolution Of The Model   53

   Risks Of Taking A Non-Literate Approach   56

   Levels Of Literacy   58

   Summary   58

   What Can You Do?   60

Chapter 3: Impact, Values and Power   63

   How We Present Impact Skews What We
   Think Counts – *Big, Shiny Endpoints*   63

   We Don't Talk About Failure (Or Harm)   65

   Who Determines What Impact Is?   67

Pressures on People And Institutions: Labour,
Visibility and Survivalism   70

Mechanising Relationships   75

Recognising Privilege   77

Towards Fairer   78

Summary   80

What Can You Do?   81

**Part 2: Eight Principles for Developing an
Impact-literate Mindset**

Principle 1: Chase Meaning Not Unicorns   85

Are Unicorns a Problem?   88

Harnessing Unicorn Energy   89

Summary   90

What Can You Do?   91

Principle 2: Work Out What Your Research *Powers Up*   93

What Can Be Mobilised?   94

Who Picks Up The Baton?   96

Partner Up …   98

… But Consider Breaking Up   100

Why? *Be an Annoying Toddler*   102

Consider *How* The Baton Passes…   104

…Before Choosing The Method   105

Prioritising (If You Have to)   107

Summary   108

What Can You Do?   108

Principle 3: Think Directionally Not Linearly   111

Why Is Thinking Directionally Useful?   112

From 'Problem' to 'Better'   112

Step 1: Describe the Baseline – 'What's the Problem'?   114

Step 2: Describe the Impact Goal(s) – If That's the
Problem, What Does Better Look Like?'   116

Summary   118

What Can You Do?   119

Principle 4: Evidence? Think *What Would Jessica Fletcher Do?'*   121

How do we Prove Impact?   123

Hard Proof: There Is No Doubt   124

Softer Proof: It's Provable When Combined   125

Proxy Measures: It Indicates But Doesn't Prove   126

Logical Proof in Uncertainty: We Can Claim If We
Eliminate All Other Explanations   127

Identifying Onward Routes; Using Events as Evidence
Waypoints   128

What Counts As Evidence Of Impact?   129

Summary   135

What Can You Do?   136

Principle 5: Create a Healthy Space   137

Why Is The Research Environment Important?   138

A Moment on Challenges and Resistance   140

Institutional Impact Literacy   142

WHY (The Purpose)   143

WHAT (The Policies)   144

HOW (The Processes)   144

WHO (The People)   145

Institutional Risks of Taking a Non-Literate Approach   146

Levels of Institutional Literacy   148

Institutional Health   149

The 5Cs of Institutional Impact Health   151

Embedding an Impact Culture   154

Summary   162

What Can You Do?   163

Principle 6: Own Your Expertise But Don't Be a Jerk   165

Summary   170

What Can You Do?   170

Principle 7: Be an Impact Lighthouse   173

What to Illuminate   174

Where and When to Shine the Light   175

Summary   181

What Can You Do?   181

Principle 8: Be You   183

Final Words   185

Frequently Asked Questions   189

Index   197

LIST OF FIGURES AND TABLES

FIGURES

Figure 1. What Is Impact?   11

Figure 2. Original Impact Literacy Diagram   54

Figure 3. Revised (Extended) Model of Impact Literacy   55

Figure 4. Impact as Up, Down or Steady   113

Figure 5. How We Make a Difference in Academia   157

Figure 6. Knowledge Mobilisation and Impact Competencies   160

TABLES

Table 1. Levels of Individual Impact Literacy   59

Table 2. Sliding Scale of Stakeholder Energies   97

Table 3. Characterising Baseline Problems and Impact Goals   117

Table 4. Common Evidence Types and What They're
         Best Suited To   131

Table 5. Levels of Institutional Impact Literacy   148

Table 6. Comparison of Healthy Versus Unhealthy Practices   153

ABOUT THE AUTHOR

Julie once got lost in a bathroom. She spends her spare time belting out 80s tunes, watching cosy murder mysteries and documentaries about Alaskan Bush people. She once ordered fake designer underwear whilst on a lot of post-surgery medication.

She is also a world expert on impact.

Vicky Williams, CEO, Emerald Group

Dr Julie Bayley is an international expert on research impact and has been immersed in the world of impact as an academic and research manager for many years. She is currently the Director of Research Impact Development at the University of Lincoln (UK) and founder and Director of the Lincoln Impact Literacy Institute – the University's strategic unit for impact – leading impact capacity building and strategy across the institution and wider university sector. She is passionate about equality and diversity within research, and is Director of Impact for the Eleanor Glanville Institute, the University of Lincoln's Strategic Unit for Equality and Diversity.

She is a regular contributor to conferences, consultations and round tables and, since 2017, she has been commissioned as Emerald Publishing's *Impact Literacy Advisor* to support their 'Real World Impact' programme. She sits on a number of Committees and Advisory Boards, including as Policy Lead on the British Psychology Society Division of Health Psychology Committee, and previously as both Director of Qualifications for the Association of Research Managers and Administrations (ARMA) UK, and Co-chair of the International Network of Research Management Societies (INORMS) Research Impact and Stakeholder

Engagement (RISE) Working Group. She collaborates nationally and internationally on knowledge mobilisation and impact, and in 2022 was awarded the *Advancing Research Impact in Society* (ARIS) Impact Innovators Award in recognition of her impact literacy work.

As well as an impact expert, she is a Chartered Health Psychologist, registered with the Health and Care Professions Council, and has been an applied researcher in behaviour change since 2003. Much of this time was spent researching sexual health, as well as healthcare staff development, public health interventions and evaluations of health and care services. Recent research has focused on improving patient-centred research, developing patient outcome measures, delivering public health evaluations and creating novel ways to review impact in funding applications and in health innovation research. Having had far too many blood clots since 2008, she is an advocate for vascular health and is a patient representative on the International Consortium for Health Outcomes Measurement (ICHOM) Central Advisory Board.

Outside of work she is a semi-lapsed double bassist, cosy mystery fan and cheese lover, and her kids describe her as having an 'adequate' face.

ACKNOWLEDGEMENTS

As with all things in life there's a lot of people to be acknowledged. But let's avoid an Oscars-style speech and say a particular thanks to some key people.

To Emerald, particularly Vicky Williams the most gloriously intelligent, capable, modest and inclusive leader a company could ever want. And she knows far too many secrets about me so I couldn't really go to another publisher. Also in Emerald, Steve Lodge for his amazing leadership of impact services, the Impact Services team for the most humbling commitment to Impact Literacy, Tony Roche for continued support and John Eggleton for constantly (yet mistakenly) assuming he's winning our childlike bickering.

There can be no acknowledgement big enough for the sheer glory that is David Phipps. A more genuine and generous person you will struggle to find, and I will never not laugh at our collectively teaching him about 'budgie smugglers'.

To the impact community in the UK, Canada, Australia, New Zealand, across Europe and everywhere else. There are too many of you to mention by name but you rock. Thanks for always saying yes to coffee and not rolling your eyes too much at random ideas. And to the world of research managers, you are amazing. Having seen brilliant leaders like Stephanie Maloney, Pilar Pousada Solino, Lorna Wilson and others, I'm reconciled to the fact I'll never grow up to be like them. But when you've been on the sharp end of Lorna's sweary Scottishness, the concept of grown up seems fairly far fetched anyway.

To Vicky, Jo and Kim for being absurdly brilliant women. I'm very sorry about the ongoing suite of accidental innuendos, but also upset you never consider these out of character. To Rebecca, Marla, Aileen and Sarah a long-term friendship from an opportunistic conference talk which makes me smile to this day. And to the Women in Academia Support Network, thanks for trying to keep this academic ship afloat for so many who are struggling.

To Mum. You're completely loopy and I love you, but I refuse to play Ludo with you until you stop elbowing me in the ribs.

To Dad. You won't read this, and you won't understand what it's about, and that's fine. Just know I did it and please stop complaining about my coffee.

To those contributed to this book, more on you later you lovely lot.

And last but certainly not least to my boys – Will, James and the husband'y Phil one. Thanks for keeping me fed with amazing bread, watered with daily coffee, and grounded with phrases like *no one understands mum's job* and *did you kill the bird with your car or with your looks?* A masterclass in family connectiveness. Love ya you weirdos.

A final thanks to anyone who's shaped this book somehow. You might see our conversations reflected in it or you might have just sparked something which took me in a certain direction.

This book is for those we bumble along with, those we love and those we want to throw pies at. It's a fascinating world isn't it?

And now, tea.

CONTRIBUTORS

There are wonderfully kind and fabulous people around, some of whom I've been lucky enough to have contribute to this book. A MASSIVE thanks to them for offering their energy, time, thoughts and comments, from whichever bit of the research and impact world they represent. Some of them were even willing to be seen with me in public to discuss things over coffee. Others less so but I'll hang around until they relent. The glorious people you'll hear from and to who I'm indebted are:

Professor Ele Belfiore is the inaugural Interdisciplinary Director for Social Inclusion and Cultural Diversity at the University of Aberdeen. She has published extensively on cultural politics and policy, cultural value and the 'social impacts' of the arts. For Palgrave, she edits the book series *New Directions in Cultural Policy Research* and she is Co-Editor-in-Chief of the journal *Cultural Trends*. She has been committed to the promotion of equality, diversity and inclusion in higher education, and a Founding Member of the Women In Academia Support Network, a trans-inclusive and intersectional charity that brings together over 13,500 members worldwide with lived experience of misogyny to facilitate gender parity and more equitable working conditions in higher education.

Dr Rebecca Brunk is a Michigander living across the pond. She is a mixed methodologist and has a background in Neuroscience and Organisational Psychology. She aims to approach every problem in academics from the angle of, how do we leverage knowledge to enact real substantial change?

Dr Gemma Derrick is an Associate Professor (Research Policy & Culture) at the Centre for Higher Education Transformations (CHET) at the University of Bristol. She relentlessly twitter-stalked

Julie and her jazz hands until she was thrilled to receive an invitation to speak at one of her seminars, and did so whilst trying to look cool. What resulted was a mutual respect and love for all things impact in an embarrassing kind of you-hang-up-no-you-hang-up kind of way. Her research focusing on the dynamics of research culture in response to external and internal reward and assessment structures. She published 'The Evaluators' Eye: Impact Assessment and Academic Peer Review' in 2018, and has been a Leading Commentator on assessment frameworks and peer review practices where she campaigns for more reflective processes of evaluation for a kinder research future. She is a Visiting Professor with OSIRIS at the University of Oslo, is on the Board for the HiddenREF, and has worked with funders, such as at the Research Council of Norway, The Academy of Finland and the Wellcome Trust about building better assessment practices for impact. She also hates writing her own bio.

Esther De Smet is a Senior Policy Advisor at the Research Department of Ghent University (BE), where she develops strategy and support on societal impact of research and research communication. She is also the Business Project lead for the Institutional Research Information System. She regularly leads workshops on communication strategy, impact, digital presence and social media. By now, she has become a valued member of the worldwide impact tribe, participating in working groups and projects, and presenting at conferences (e.g. INORMS, EARMA, AESIS, etc.).

Dr Kieran Fenby-Hulse is an Experienced Lecturer, Facilitator and Researcher with over 15 years of experience in research, strategy development and organisation development. He takes an interdisciplinary approach to both research and teaching, drawing on artistic techniques and practices to explore, challenge and interrogate notions of responsible and shared leadership, continuing professional development, and communication and engagement. He is the Co-founder and Managing Editor for the *Journal of Research Management and Administration*, a Reviewer for the *International Journal of Doctoral Studies* and Member of Association for Research Managers' EDI Advisory Group.

Dr Elizabeth Gadd is a Research Policy Manager at Loughborough University. She chairs the International Network of Research Management Societies (INORMS) Research Evaluation Group and Co-champions the UK Association of Research Managers and Administrators (ARMA) Research Evaluation Special Interest Group. She founded the LIS-Bibliometrics Forum and The Bibliomagician Blog which provides bibliometric advice and guidance 'by practitioners, for practitioners'. She was the recipient of the 2020 INORMS Award for Excellence in Research Management Leadership.

Dr Tamika Heiden is the Founder and Head Inspirer of the *Research Impact Academy*. She previously worked in Health Research and Research Coordination for more than 15 years and for the last eight years has been Consulting to researchers, funders and research organisations across the globe. Her background in knowledge translation and research impact, along with her dedication to improving social outcomes from research, led her to develop the Research Impact Academy. Her goal and purpose is to ensure that research is relevant and accessible to the people who need it (www.researchimpactacademy.com).

Dr Chris Hewson studied Social and Political Sciences at Cambridge, and took his PhD in Sociology at Lancaster University. His PhD examined the development of community radio and television services in the face of social and regulatory change. He stepped into the newly created role of Social Sciences Research Impact Manager in March 2017, having previously worked in impact support roles at the Universities of Manchester and Salford. He provides Expert Assistance to York's Social Sciences Departments regarding all aspects of research impact, including but not limited to funding, policy engagement and the Research Excellence Framework (REF). He manages York's Economic and Social Research Council Impact Accelerator Account. He is always keen to hear from organisations and groups interested in collaborating with the University's social science researchers.

Helen Lau is currently the Associate Director of Knowledge Exchange at Coventry University. Having worked in research commercialisation, knowledge exchange and innovation for 17 years across roles at universities, regional development and SMEs. She is passionate about research and university innovation making a difference and changing people's lives for the better, essentially linking research impact and Knowledge Exchange (KE) together to try and make the world a better place for everyone. She is an Institute of Directors Ambassador for Inclusion and Diversity and an Active Non-executive Director with smaller companies, spin outs and charities. She's a full-time working Mum of two and loves sharing her passion for innovation, impact, KE and inclusivity with everyone and anyone

Dr Kellyn Lee is a BPS Chartered Psychologist, Academic and Founder of www.materialcitizenship.com and www.dementiac-arehub.co.uk. She works with the social care sector to improve the lives of people living with a dementia and those who care for them. She also works as a Project Officer of the NIHR ENa-bling Resarch In Care Homes (ENRICH) project via the London School of Economics and Political Science. More details on her translation of research into practice is available at https://www.theguardian.com/society/2021/jan/14/everyday-objects-people-with-dementia-quality-of-life-care-homes and https://www.you-tube.com/watch?v=1JAP_iYtHtQ

Dr Stephanie Maloney is Director of Research and Enterprise at the University of Lincoln and supports the institution in establishing a research and enterprise culture. She leads a department responsible for research grant support; consultancy and educational contracts, business start-up & growth; IP commercialisation; funding partner-ships; research environment; ethics, governance & integrity; research data & systems; knowledge exchange and support for regional busi-nesses. Prior to joining the University of Lincoln, she worked at the University of Birmingham and was responsible for facilitating projects between the University and industry, especially with public funding. She holds PhD in Oncology from the Cancer Research UK Centre at the University of Birmingham. External to the University, she is a Member of UKRI's Research Organisation Consultation Group advising UKRI on all aspects of research policy, process and procedure from the research organisation perspective. She is engaged in driving forward regional innovation and growth through, for example, South Lincolnshire Food Enterprise Zone, Greater Lincolnshire LEP (GLLEP), Lincolnshire Growth Hub, GLLEP Innovation Council and Lincoln Science & Innovation Park.

Wilfred Mijnhardt is Policy Director at Rotterdam School of Management, Erasmus University. He brings over 25 years of experience in Research Policy Development and Institutional Advancement. He is passionate for universities, business schools, responsible research and education, excellence and impact. As Executive Director of Erasmus Research Institute of Management (ERIM) (till 2014), he has been a Pioneer in organising for academic development and the impact support organisation, with special expertise on quality assurance, productivity, academic and societal impact of research and the renewal of faculty management. In his current role, his energy focuses on the strategic transition of universities and business schools towards an impact-driven mindset. Internationally, he is an Active Member in networks like RRBM, PRME, AACSB, EFMD, amongst others. He holds a bachelor degree in Economics, a master's degree in Public Administration and a postgraduate diploma in Management of Change,

Dr David Phipps is the Administrative Lead for all research programmes and their impacts on local and global communities at York University (Toronto, Canada). He has received honours and awards from the Canadian Association of Research Administrators, Society for Research Administration International, Institute for Knowledge Mobilization, International Network of Research Management Societies and the EU-based Knowledge Economy Network. He received the Queen Elizabeth II Diamond Jubilee Medal for his work in knowledge mobilisation and research impact and was named the most influential knowledge mobiliser in Canada. He sits on knowledge mobilisation committees around the world and is Network Director for Research Impact Canada.

Derek Stewart is a Former Teacher treated successfully for throat cancer in 1995 by clinicians who were active in research meant he didn't necessarily receive the 'best evidenced based treatment' but he still has a voice. He is a Strategic Advisor with the HRB Trials Methodology Research Network, Ireland and Hon Professor at the University of Galway.

Dr Mark Taylor, according to his last consultant's letter, has 'relapsing remitting multiple sclerosis, a sacral nerve sheath tumour, hips tendinitis and labral tear, knee meniscal injury, shoulder adhesive capsulitis and now tentative degenerative arthritis'. He currently works part time at the National Institute for Health and Social Care (NIHR) but is a patient advocate for multiple sclerosis (https://www.youtube.com/watch?v=823gl0-74-o) and other broader health issues (https://blogs.bmj.com/bmj/2018/09/13/mark-taylor-does-it-matter-that-letters-between-consultants-and-gps-are-not-addressed-to-patients/).

Vicky Williams is Chief Executive Officer of the Emerald Group, which comprises Emerald Publishing and Emerald Works. She has held a variety of senior roles during her 20+ years in the publishing sector, across editorial, business development, product development, marketing, digital and HR. She is an International Advisory Board Member for the University of Bradford, and an Advisory Board Member for the University of Lincoln's Impact Literacy Institute. She also works within the local community on Bradford's Economic Recovery Board and is a Trustee of the Keith Howard Foundation. Both in and out of work, she is a Keen Advocate for gender diversity, having launched Emerald's Equality, Diversity and Inclusion programme in 2016, and speaks widely on this topic at global forums and events.

Lorna Wilson has worked in Higher Education for just over 10 years and is currently Co-Director of Research and Innovation Services at Durham University, and Chair Elect of the Association of Research Managers and Administrators (ARMA). She is a HUGE Research Geek, with a background mainly in research development and funding. Her current role involves her leading on various areas for her institution including research development, operations, culture and strategic projects. When she's not cheerleading research she's a mum of two black Labradors in the Toon with her husband.

Professor Clare Wood is Professor of Psychology at Nottingham Trent University, and describes herself as a 'research mongrel' in as much as her work positions her across disciplines rather than within her own. She is also a 'research magpie', insofar as she is drawn to shiny things that bring her joy when it comes to particular projects and people. She is interested in children's literacy development, children's voice and rights, and the use of technology to support learning. She drinks more tea than any human bladder should actually be able to cope with and fondles yarn in her spare time. Her daughter is embarrassed by her.

Introduction

Julie Bayley

Well hello.

Welcome to this book on developing an impact literate mindset. Its aim is to help you understand impact, think about how to 'be you' in impact, and develop a set of guiding principles for impact in practice. You might be starting from scratch, need a refresher or want to somehow reset your thinking as you approach impact further on in your career. Welcome all.

Impact is not new, nor is it – unless it actually is – rocket science. And at its heart, a very simple principle – to make a positive difference through research. That's it. Honestly. Impact is when people outside of academia benefit from the research we do within it. There are all sorts of national agendas and strategies that demand impact be delivered at scale, in accordance with certain rules and corroborated to a certain level of proof, but that doesn't detract from the gloriously simple premise that it's about making the world a bit better. Everything else pretty much is smoke and mirrors, and it can be easy to get lost in the haze. I don't mean that it doesn't take effort to understand, but rather it isn't something too niche or complicated to get a handle on.

This book aims to cut through any mists, giving you chance to understand what impact is and align your impact sat-nav. It's about celebrating the opportunity we have within the research world to make a difference, whilst also recognising the challenges of doing impact in already pressured environments. And it's about ensuring that research of any size and from any discipline has a fair line of sight to impact, whilst acknowledging that even the best laid plans can fall foul to the negative mood created by having no biscuits at a meeting. Impact literacy is simply the ability to understand what impact is, at a level which allows you to not just do impact but get under its bonnet and see how its engine works. And an impact literate mindset is about building that understanding into your thinking, your judgements and your actions across your work and the environments in which you work.

Perhaps the biggest reason this book exists is that I hate how battered and bruised people can feel by impact. Impact can feel exhausting, because research can be exhausting, and because academia can be exhausting. If you add commuting, kids, health issues, caring responsibilities, job precarity, pandemics and all else into the mix, there's a less than healthy correlation between life and the need for a cash-and-carry sized bar of Dairy Milk. Sometimes the prospect of giving energy to something else when you've already put all available lifeforce into the work–life–chocolate trinity can feel overwhelming. I'm certainly not going to patronisingly suggest that workload is simply overcome by a change in thinking, but I absolutely believe that getting to grips with what impact actually is – *impact literacy* – makes the whole thing feel less overwhelming, more under our control and far more resistant to unreasonable expectations. As someone who still believes strongly that research impact is a genuinely good thing, having been battered and bruised by some of the pressures to do it, I keep coming back to the simple point that the opportunity to help society with our research is the right thing to do. It matters. And it matters enough to keep pushing for better, healthier and fairer ways to do it. So this book is written from me to you, to help keep the faith when it all goes a bit pear shaped.

I want to be clear from the outset that it doesn't aim to provide a prescriptive reference-heavy framework or stepwise 'how to' for impact. In fact, if you're looking for unequivocal, data enforced arguments about which logic model is better, then you'll be a teensy bit (very) disappointed. There are already many fantastic resources out there providing models, frameworks, public engagement good practice and the many other facets of impact. I don't write this suggesting that this book overrides or replaces them – quite the opposite. The wealth of available insights is incredible, but particularly for those stepping newly into impact it can be daunting to know how to dive into something so vast. My aim is to help you bring values and principles to the start of your impact thinking to *then be able to* draw on these insights. If tools and frameworks are the recipe to help you make an impact cake, this book helps you think about why you need to make the cake, why *that* cake, for what occasion, who's going to want to eat it, who might be allergic and how you can avoid dropping it on the floor. My stance is this: you can't make full or effective use of the tools available if you don't have your impact head screwed on first.

The content comes from many years of experience being right in the middle of this thing called impact as an impact lead, applied researcher, research manager, psychologist, impact consultant, patient, mum, carer and general human. It comes from seeing people embrace/love/be suspicious of/hate impact and all else in between. Whilst much of my lens on impact is shaped by the UK context, it's not based on the UK experience alone. Nor is this book about the UK, about any specific agenda, or limited to my reflections. My aim is just to reflect the impact world I see, help you reflect on yours, give you some ways to get a mental foothold on the topic, and try to answer any stupid (but categorically aren't stupid) questions you might have. And its tone is, well, me, the same as if you were listening to me rattle on at a conference or in a workshop. Apologies, you may need wine.

At the broadest level this book is for anyone within academia interested in impact, but it will be most use to researchers, impact managers, those involved in knowledge brokerage or in institutional roles which support impact in some way. I also hope it's useful for those entering academia and needing to make sense of this odd impact shaped thing. I don't start with any assumption of where you are on the impact experience spectrum, but instead hope it helps you build healthy approaches to impact whatever your starting point. If you are one of those people who loves theory or gets excited about a significant statistical result (just so we're clear, that isn't me – in the same way that I don't understand nanotechnology or Love Island), then fabulous. We need you – your work forms the foundations of research and signals the spark of an impact possibility. But if you are one of those people who gets their joy from connecting with people or getting your hands dirty implementing things in practice, we need you too, to energise the work into life. There is a place for everybody and no template for what counts. Grab a coffee and pull up a chair.

Throughout this book I'll use phrases like 'your research' and 'your impact' as a semantic shortcut to avoid constantly repeating 'your impact, or that of the research you support, or you help communications on, or are brokering into society, or are commissioning ....'. The possible angles on impact are endless, so to simplify the narrative I will say 'your research', with the absolutely expressed wish that you translate this and the examples into the perspective that suits you.

STRUCTURE OF THE BOOK

Impact is a difficult topic to cover generically to suit everyone, so by necessity I'll need to use a range of illustrative examples rather than cover every possible permutation. And I'll very unprofessionally use a lot of analogies and references to film, TV and other random things. This is because I have a simple brain and analogies are how I make sense of the world, but also because I hope these give you some much clearer visual shortcuts for understanding what can feel an unmanageable beast. If some of my references don't land – entirely possible as the process of writing this book has made it clear to me how eclectic my internal library is – don't worry. They're used to illustrate, not explain, and the fuller text will explain what you need to know. And if you can't find your exact research represented, remember that much of what I'm using is examples to help you reflect on your own context.

Each chapter contains thoughts on what you can do next, with questions to help steer you in the right direction, and there are comments, tips and reflections running throughout from a range of fabulous people in and around research. These include impact experts, funders, publishers, academics, research managers, knowledge exchange experts and patients, each of whom has first-hand understanding of how impact intersects with other areas of life, and how we can do it well. You'll see their comments throughout the book, and I am indebted to them for their contributions.

The book is split into two parts:

Part 1 focuses on impact within the research landscape, and how an impact literacy approach can help.

Chapter 1 is the most 'textbook' part of the whole thing, covering the basics of impact, what it is and isn't, how it's defined, what drives the sector to pursue societal change and how it all works in practice. Chapter 2 delves more deeply into impact literacy, exploring what impact literacy is, what it means to be literate and why the ability to critically judge aspects of impact is so important. Chapter 3 then looks into some of the pressures and challenges of impact which bring values very much into the frame.

In *Part 2* we change focus and look at a set of principles for approaching impact in a meaningful, literate way:

- Principle 1 – Chase Meaning Not Unicorns – focuses on ensuring we base our activities on what matters, not what is most impressive.
- Principle 2 – Work Out What Your Research Powers Up – covers how research has a chance to create impact every time the 'baton is passed'.

- Principle 3 – Think Directionally Rather Than Linearly – looks at how impact can be best thought of as changes in various directions, rather than linear paths
- Principle 4 – Evidence? Think – What Would Jessica Fletcher Do? – focuses on how we can prove impact, either by finding the smoking gun or assembling a case.
- Principle 5 – Create a Healthy Space – shifts gear to think about how we, as individuals, can help build healthier approaches to impact in our institutions, departments or groups.
- Principle 6 – Own Your Expertise But Don't Be a Jerk – jointly covers how to overcome imposter syndrome and/or not be a complete pain in the backside.
- Principle 7 – Be an Impact Lighthouse – looks at how we can integrate impact literacy and healthy practices across the various aspects of academic life.
- Principle 8 – Be You – reminds you to be authentic as you navigate this world of research translation.

At the end of the book there is a 'Frequently Asked Questions' section, which tries to condense some of the points above into specific queries.

This book won't change your life, unless your main problem is a wonky table needing a book-sized wedge, but it will help you equip yourself with the thinking to approach impact well. You'll most definitely still learn far beyond this book, have wins, make mistakes, judge, misjudge, celebrate, feel sucker punched, feel elated and everything else, but knowing you're starting with healthy values has got to be the right way to embark on the journey.

My aim is to leave you with some principles to anchor yourself in this thing called impact.

My hope is that you find your impact mojo and feel empowered to do it. Happy reading

Julie x

Part 1

IMPACT, IMPACT LITERACY AND VALUES

Chapter 1

WHAT IS RESEARCH IMPACT?

*Research impact is the provable effects (benefits) of
research in the 'real world'.*

That's basically it. That's what impact is. Thank you for reading,
have a great day.

Ok so it's a bit more nuanced than that, but that's the heart of
it. Research impact (from hereon just 'impact') is the term which
describes how things outside of academia are somehow different
because of research. It's about using research to make a difference
to the world in which we live, be that to people, the environment,
culture, the economy or any other facet of life.

WHAT IMPACT IS

There are various definitions of impact in operation around the
sector, including:

> '*the demonstrable contribution that excellent research
> makes to society and the economy.*'
>
> UK Research and Innovation[1]

---

[1] https://www.ukri.org/councils/esrc/impact-toolkit-for-economic-and-social-sciences/defining-impact/

'an effect on, change or benefit to the economy, society, culture, public policy or services, health, the environment or quality of life, beyond academia.'

UK Research Excellence Framework,[2] 2021

'The contribution that research makes to the economy, society, environment or culture, beyond the contribution to academic research.'[3]

Australian Research Council

[Broader impacts are] *'the societal impact of the proposal and may be accomplished through the research itself, through the activities that are directly related to specific research projects, or through activities that are supported by, but are complementary to the project.'[4]*

Broader Impacts of the National Science Foundation

These definitions all centre on the realised potential of research to make a difference outside of academia. In short, they align on impact being the *provable effects (benefits) of research in the 'real world'* (Bayley & Phipps, 2019), which I've massively overly simplified in Fig. 1.

Before we go any further, let me make two points about Fig. 1. Firstly, you'll notice that the research 'bubble' sits within the bigger 'world' one. That's because, unless I'm very much mistaken, universities and research institutes are *part of society* rather than some ethereal, existential deity looking objectively and cleverly from on high (*Disclaimer: we all know people who believe the contrary*). Secondly, the separation between research and society in Fig. 1 *only* relates to where the benefits are felt. It does not and NEVER SHOULD be taken as a suggestion that the research process itself should be separate from society. Indeed it's very much the opposite; collaborative, coproduced and participatory research processes offer some of the strongest pathways to impact. So much so that

---

[2]Research Excellence Framework 2021 Guidance on submissions. https://www.ref.ac.uk/media/1447/ref-2019_01-guidance-on-submissions.pdf

[3]Australian Research Council Research Impact Principles and Framework. https://www.arc.gov.au/about-arc/strategies/research-impact-principles-and-framework

[4]Advancing Research Impact in Society Broader Impacts. https://researchinsociety.org/broader-impacts/

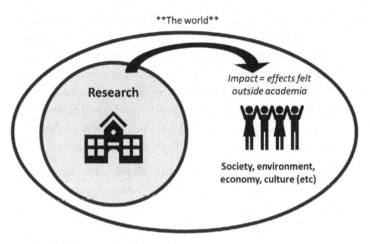

Fig. 1. What Is Impact?

I'm not going to itemise coproduction as a separate path to impact, but rather underline its importance in everything we seek to do in impact. Coproduce; *message ends.*

Back to the definition of impact. Impact is the changes we can see (demonstrate, measure and capture), beyond academia (in society, economy, environment or elsewhere) which happen because of our research (caused by, contributed to and attributable to). Impact may look and operate slightly differently across disciplines, and for fundamental versus applied research, but ultimately is about connecting academic research to changes in the real world.

Let's break this definition down further:

Provable: We'll talk about evidence later (Principle 4), but we must be able to see, demonstrate, measure, corroborate, substantiate or otherwise prove any claims of impact. Otherwise how can we know we're not just assuming we've made a difference?

Effects/benefits: Impact is the state of something outside of academia being different because of our research, ideally better because we don't really want to make things worse, although we'll discuss negative impacts in Chapter 3.

Of (from) research: For it to be research impact, any effects we claim must arise in some way from our research, *as distinct from* other aspects of academic life such as teaching or consultancy alone. Note: With research having many types and flavours, the internationally recognised definition of research comes from the Frascati Manual[5] which describes research as 'creative work undertaken on a systematic basis in order to increase the stock of knowledge' conducted as basic research, applied research or experimental development. Research impact needs to extend from this new knowledge, even if it's then applied through actions, such as consultancy.

'Real world': that is, the effects are felt outside of academia, for example, in the economy, the environment, culture, health, commerce, manufacturing, shopping, dog walking, viola playing or the enjoyment of clowns. This is not an exhaustive list. We won't go into this in any specific chapter, because, well, it is what it is.

I need to momentarily sound a loud flashy siren to comment on the fact that the precise endpoints of these aspects can and do flex in different agendas. Funders, assessment, national missions and any other formalisation of impact can place exact boundaries at different places, meaning that 'what counts' for one thing might not *quite* count for another. Context is everything when it comes to formalities, and rules will always vary, but however it's operationalised 'impact' remains anchored to real-world change. Whatever debates we have inside academia, and wherever we put the lines in the sand, it's not unreasonable that society expects us to try and use our research to try and make the world a bit better.

Back to what impact is. The principle that impact equals research-led differences in the *real world* is the single most important point to absorb in this book. And yes, 'real world' is a loaded term. I use it for blunt simplicity to mark a clear line between academic research – that glorious world of theoretical, conceptual, knowledge-rich, self-referent, peer-judged, paradigm-shifting, reproducible, confirmatory and illuminating world of exploration – and the part of life where we visit the doctor, recycle, buy a dog or pretend to understand those films that win Oscars. If research

---

[5] https://read.oecd-ilibrary.org/science-and-technology/frascati-manual-2015_9789264239012-en#page1

offers new knowledge, impact is the expression for the conversion of that into change in the wider world. For that reason I'm going to deliberately shorthand non-academia as 'the real world' from hereon in to reinforce the need to think beyond the creation of, or interest alone, in knowledge. And if you choose to extend that analogy to suggest academics aren't real people, that's on you.

WHAT IMPACT ISN'T

The basic principle of impact is joyously simple; it's the actions around it which are far more involved. Let's face it, there is something immediately in conflict about taking a piece of fancy academic research and trying to put it into usable societal practice. Like taking a lute to a heavy metal gig. One of the things that reaaaaaaally (*extra vowels intentionally added for emphasis*) doesn't help is the mental image conjured by the word 'impact'. Dictionary[6] definitions of impact talk of 'striking', 'forceful contact' and 'collision'. We get a sense of a meteoric crash, or wrecking ball, and even Miley Cyrus would struggle to make that seem like a good thing. Thankfully as soon as you start to unpack the term, it becomes clear that impact is far less of a crater and far more glitter cannon, but it does mean we end up having to undo presumptions before we can really get going.

To fully clarify what impact is, it's also therefore extremely important to be clear what impact *isn't*. Only that way can we get some firmer boundaries to guide our thinking. If impact is change in the real world, of any size or shape, it therefore is *not*:

Dissemination, that is, the act of communicating research. Vital for making research visible, but not impact until the research is used.

Academic citations or other bibliometrics. These measure academic attention to an output, rather than showing or providing a measure of real-world change.

---

[6] https://www.dictionary.com/

Measured by markers of attention, such as social media followers or retweets as whilst these may give a metric of people noticing your communications, they don't themselves show *change.*

Measured by markers of academic reputation, such as invited keynotes and editorials. Fantastic for our careers, but aren't change outside of academia.

Engagement alone. Connecting research with society is massively important for impact, it just *isn't the same as* impact.

Big by definition

Let me round off this section with two final points about how the boundaries of impact fit with the wider life of academia.

1. The inclusion of *research* in the definition of impact doesn't suggest that changes arising from research are somehow superior to benefits arising from other things we do in academia. We need only speak to colleagues to hear the positive stories of teaching and learning, performances, alumni success, entrepreneurial incubation, activities to raise the aspirations of children (and so on) to know that research is just one part of the academic tapestry. It's just that research offers a unique contribution to the world, and so it's incumbent on us to maximise the benefits it can have.

2. There is no suggestion that impact is superior to research itself (or by extension that research without impact isn't valuable). Research and impact just aren't the same thing. Locking in the term impact to mean benefit *outside* the research sphere simply allows us to distinguish changes outside of academia from those within it and value them each for their own merits.

TYPES OF IMPACT

Back to what impact is. The term impact relates to real-world change, and as there's a gazillion ways the world can look different, there is no single type of effect nor exhaustive list of every kind of impact.

That would be insane. Trust me, I've tried. So instead we find ways to categorise, compartmentalise or otherwise 'chunk' what impact is to get order on something which is otherwise potentially – and excuse the technical term here – massive.

It's pretty common for impact to be categorised by the context in which it hits, for example, social impact (changes for society), economic impact (changes to the economy), environmental impact (changes for the environment) and so on. You see this language fairly regularly in formal definitions about impact, and fuller expressions of what funders and other governing agencies view as impact. Let's call these *impact domains*. What's brilliant about impact domains is that they can be in the same field as the research – social research having social impact, environmental research having environmental impact, etc. – or in a completely different domain altogether. Economic research could have social impact; arts research could have environmental benefits. The impact world is your oyster in that respect.

However, knowing where research benefits can land is not quite the same as understanding *the nature of the thing that changes*. And it's here that the UK's Economic and Social Research Council offer an immensely useful three part categorisation[7]:

> Instrumental impacts – *a change in a thing*. These are perhaps the most easily brought to mind types of effect, tangible benefits where there is a new or changed 'thing'. Common examples include new services, new policies, changes in behaviour, changes in visitor numbers, profits, new audiences, new treatments (etc.). Instrumental impacts can be summarised by the prefix – *there now is*.

> Capacity building impacts – *a change in what can be done*. These are benefits which reflect the new or increased ability to do something because of research. Common examples include increased skills, opportunities afforded by technological advancement, or new connections, networks and partnerships. Capacity building impacts can be most easily expressed by the prefix – *now able to*.

---

[7] https://esrc.ukri.org/research/impact-toolkit/what-is-impact/

Conceptual impacts – *a change in how something is thought about.* These are changes on understanding, awareness, attention, perception, the tone of debates or any other effect on how things are conceived. Conceptual impacts can be harder to measure because they are the more abstract of the three types, but can be seen in (for instance) topics newly bought into parliamentary debates, the change in the way something is presented in the media, or reports of people feeling less stigmatised. Conceptual impacts can be most easily expressed by the suffix – *is now thought about differently.*

Taken together these give us a sense that research can yield changes in certain aspects of life (domain), by changing a thing, a way of thinking or the ability to do something (type). To complete the picture, we then need to recognise that benefits can arise from any part of our research:

Impacts arising from findings: new knowledge somehow leading to change.

Impacts arising from the research process: the practices of, or involvement with research that can be a catalyst for change themselves.

By combining these you can start to more clearly articulate – mentally and on paper – the benefits your research can have:

As a type (instrumental, capacity or conceptual)

+

In a domain (society, the economy, in business, the environment, etc.)

+

Arising from findings or process (because of findings or because of process)

In Principle 3, we'll delve into this further, particularly thinking about impact as a *direction* of change.

PROVING IMPACT

If impact is the provable benefits of research in the real world, then it's necessary to explore what's meant by proof. Some formal agendas have a requirement to submit substantiating evidence of any impact claims so that they can be verified and accepted. Arguably less forensic in the veracity of evidence need, similar requirements are seen within funders who are increasingly requiring researchers to deposit evidence of outputs, outcomes and impacts into databases. Whatever formal requirements, the key thing to focus on is the need to determine how claims can be reasonably substantiated.

We will talk about impact evidence in more detail in Principle 4, but in essence proving impact comes down to five questions which may be answered in this or any more iterative order:

1. *What changes?* that is, what is the impact?

2. *Who can tell you?* that is, who would know? Who has eyes on what's happening?

3. *How can you demonstrate it?* What kinds of indicators or markers of change (qualitative or quantitative) could you look for?

4. *Where can you get it from?* Where can data, information or feedback on these changes be found? What are your sources?

5. *What format?* In what way could you collect it?

WHY DO WE DO IMPACT?

I've intentionally left this part until we've defined impact, and kept it as an overview, so it contextualises rather than leads the explanation of what impact is.

Applying research to social, economic, environmental and political problems is far from new. It's a longstanding feature of academic practice, sometimes described as the 'Third Mission'[8] of universities alongside teaching and research. Many areas of research activity – which is often publicly funded – strike right at the heart of social need, and there are numerous examples of how research has helped society. You only need look at the rapid development of the COVID vaccine to see how pivotal research is in certain areas of

---

[8] For example, see UNESCO Digital Library https://unesdoc.unesco.org/ark:/48223/pf0000157815

our lives. But outside this unprecedented and urgent need for rapid research evidence, there is often appetite for, yet a far more scenic route to, research application.

Impact sits amidst a range of agendas relating to research accessibility, transparency and democratisation of knowledge. Open Access and broader Open Science efforts seek to reduce barriers to journals, books, data sets and other products of research by eliminating paywalls, unblinding reviews and creating platforms for reuse and reproducibility. Public involvement initiatives, particularly exemplified through public and patient involvement in health research, once a call to arms are now more frequently mandated as part of the research development process.

The parallel drive to translate research more effectively into society has been amplified and accelerated by the introduction of more formal impact agendas, particularly around funding and assessment. This has happened in different ways and at different paces globally, with some countries still determining how, or if to include impact in expectations of academia. Whilst still hugely varied in the extent to which it's instrumentalised in research, impact in most formally enshrined in funding and research assessment, with more social and individual missions underpinning motivations for many. Let's go through some key drivers of impact: funding, assessment missions and personal motivation.

Impact in Funding

Impact in funding can be characterised by the question: *In relation to* [the funder's] *goals, what real-world benefits – or traction towards them – will this research achieve?*

Impact is an increasing important aspect for many funders around the world. Within the UK, for example (and recognising entirely there are many different models internationally), research impact has been substantially driven through its introduction to both arms of the government's dual funding system.[9] This system operates by providing

[9] See https://www.gov.uk/government/publications/dual-funding-structure-for-research-in-the-uk-research-council-and-funding-council-allocation-methods-and-impact-pathways

research investment through two major channels: competitive funding programmes run through subject-focused research councils, and quality research (QR) income which is directly allocated to universities on the basis of achievements in research assessment exercises. For the former, many UK funders mandate robust impact plans at the outset of a research programme, connected to their strategic aims, and requiring of monitoring throughout and beyond the project lifetime. For many years (until 2020) these plans were encased in a separate 'Pathways to Impact' section of the bid, wherein applicants provided a short account of their impact plans, goals and beneficiaries such to convey their legacy thinking to peer reviewers. Whilst this has section has since been removed the requirement to plan impact still remains, with an expectation it will now be weaved throughout the bid instead. We'll come to the latter under 'assessment' below.

Impact is not restricted to Research Councils, nor to the UK. The US National Science Foundation has similar requirements for establishing the potential of research to have impact, and in Canada, most academic research funding agencies require a strategy for knowledge mobilisation, knowledge translation and/or commercialisation, depending on the funder. Elsewhere in the UK, major funders such as the National Institute for Health Research (NIHR) require impact plans, and provide a number of mechanisms to reinforce impact potential including access to expert guidance during the application process. Beyond these examples, it is also common for impact to reflect funders' missions, particularly where they relate to societal issues such as health, education, culture, economic growth, environment or other area of need.

Requirements for impact vary by geography, funder, funder goals and the translational remit (or not) of work to be funded. Where impact is not (or not yet) a significant part of the research landscape, funders may not (or not yet) build it into the formal bidding process. Where schemes relate to application, translation or other use of the research, it is very typical for funders to require impact plans; where schemes are for more theoretical or person based (e.g. fellowships) there may or may not be a requirement to give a 'line of sight' to impact. Ultimately there are no set or harmonised rules about funders' requirements, but there has been and continues to be an increasing visibility of impact in the bidding process, cementing impact as an ever more significant part of the research landscape.

'The European Commission's policies and funding programmes are driving a lot of the attitudes and responses
to impact in mainland Europe. We have SDG's and missions running through the Horizon Europe programme,
technology readiness levels and very prescriptive impact
goals, extensive templates for communication/dissemina-
tion/exploitation (apparently these are all different things)
sporting lists of key performance indicators (KPI's), and
results platforms and impact prizes, etc. Researchers and
research managers are struggling to keep up with the huge
amount of instructions, briefs and position papers – we
like ambition but how do we keep it manageable? Dealing
with this does not only require building up impact literacy
among the research community but also within the research
management and administration (RMA) community. And
beyond that, pushing for impact literacy among EU policy
makers so that they are managed in their expectations and
will support an impact agenda which is healthy for all.'

Esther De Smet, Senior Policy Advisor,
Research Department of Ghent University

Impact in Assessment

Impact assessment can be characterised by the question: *What real-
world benefits did this research achieve?*

If impact in funding reflects the need to plan impact, assessment
led initiatives reflect the need to demonstrate actual achievements.
Impact assessment is an increasingly common feature of research,
typically undertaken to identify the contribution of research to
the funder or the wider economy ('return on investment') and/or
allocate funds for further research.[10] Not all countries have impact
assessment processes, although it's clear the potential is being

---

[10] See Adam, P., Ovseiko, P. V., Grant, J., Graham, K. E. A., Boukhris, O. F.,
Dowd, A.-M., Balling, G. V., Christensen, R. N., Pollitt, A., Taylor, M., Sued,
O., Hinrichs-Krapels, S., Solans-Domènech, M., Chorzempa, H., & for the
International School on Research Impact Assessment (ISRIA). (2018). ISRIA
statement: Ten-point guidelines for an effective process of research impact
assessment. *Health Research Policy System*, 16, 8. https://doi.org/10.1186/
s12961-018-0281-5

explored in various territories. And whatever I might say as the current state of play in assessment will immediately be out of date, such is the potential pace of change. So I'll keep this about the bigger picture rather than the detail, and use illustrations to express what impact assessment usually seeks to do, and how that is operationalised in practice.

The UK's REF assessment has been somewhat the poster child for major impact assessment. Many of you will know about REF, but just as a very quick headline for those of you who aren't already twitching ... by way of quick history, since 1986, cyclical assessments of the quality of UK research quality have underpinned the process of QR allocation (as per the dual funding model above). Occurring approximately every seven years, in a significant change for 2014, impact was newly introduced to form 20% of the assessment weighting alongside traditional measures of outputs and contextual data on the research environment. With impact judged to have been a successful addition, it was retained in – and increased to 25% weighting – for the 2021 cycle. For REF 2014 and 2021, institutions had to develop a quota of impact case studies within broad subject areas or 'Units of Assessment'. These case studies consisted of a narrative account of the research undertaken at the institution, by staff employed at the institution, in the 20 years (approximately) prior to the final submission, with the necessary references to demonstrate the research was of a suitable ('2 Star') quality. The case study then describes the impact(s) arising from this research, within the (approximately) last seven years of the 20-year timeframe, with all claims corroborated by a body of evidence. The number of case studies required was a quota based on the number of staff entered into the overall submission (if you like rabbit warrens, feel free to check out all the rules about this, otherwise just go with it), with the cases then assessed by a panel of academic and non-academic experts within the remit of the Unit of Assessment. These assessments yielded a star rating between 1 ('Recognised but modest impacts in terms of their reach and significance') and 4 ('Outstanding impacts in terms of their reach and significance'), or an 'Unclassified' rating in unfortunate cases judged ineligible or without impact, with funding allocated proportionately to the higher scoring cases. And breathe ....

REF of course is not the only form of impact assessment, the only model of national assessment, nor are national schemes the only size. Impact can be assessed by funders, charities, institutions, professional associations and many other entities. The precise scope and eligibility criteria – or 'what counts' – for assessment will vary by the agency administering the process. Criteria commonly include impact within certain timeframes (particularly where assessment is cyclical), restricted to certain groups within academia (e.g. including or not including impact from students' work), and the extent to which the assessment credits impact-related elements such as engagement or academic influence. It is common for such assessments to use narrative approaches, which lend themselves to peer review, but with an appetite for metrics where they can be reasonably provided to add to the story. Whatever the specific rules, and whatever your view of impact assessment, it's important to acknowledge its role in the research ecosystem.

> 'The REF serves as a bogeyman for many, with impact case studies often viewed as selective, overly engineered and suspiciously linear. However, it's important to not throw the baby out with the bath water. REF impact has not only opened up a wider debate within UK academia, it has also encouraged universities to invest in impact support structures, often for the first time. Moreover, not withstanding the ever present dangers of instrumentalism, the strategies required to achieve high quality evidenced REF impact largely coincide with accepted good practice.'
> Dr Chris Hewson, Faculty Research Impact Manager (Social Sciences), University of York

Impact in Missions

Impact in missions can be characterised by the question: *How can the world be a better place, and how will/did the research contribute?* For institutions it can also be characterised by the question *'what is our vision of change?'*

The principle that research can improve the world is a fundamental driver for many groups, and whilst funding and assessment can kickstart us into thinking about the impact of our research, missions give us a clear sense of what change could look like on a bigger scale.

The United Nation's Sustainable Development Goals[11] (SDGs) provide perhaps the most prominent example of a 'mission' to which research can help solve societal problems. The SDGs are a set of 17 connected, global goals forming a 'shared blueprint for peace and prosperity for people and the planet, now and into the future':

1. No poverty

2. Zero hunger

3. Good health and wellbeing

4. Quality education

5. Gender equality

6. Clean water and sanitation

7. Affordable and clean energy

8. Decent work and economic growth

9. Industry, innovation and infrastructure

10. Reduced inequalities

11. Sustainable cities and communities

12. Responsible consumption and production

13. Climate action

14. Life below water

15. Life on land

16. Peace, justice and strong institutions

17. Partnerships for the goals

---

[11] See https://sdgs.un.org/goals

Each goal has a set of specific targets with indicators by which change can be monitored. Many funding opportunities align to delivery of these goals, reaffirming the importance of research in achieving change. Whilst research is not the only way SDGs are achieved, it does offer a significant and unique contribution.

Missions though are not only enshrined in mega agendas such as SDGs. Funders may set a particular 'vision' for change, charities routinely have a clear mission (to support, e.g., a specific health condition, social issue, endangered animal or historic venue), and commercial organisations, professional associations, public sector institutions and many others have clear goals for what change they want to drive. Indeed universities themselves commonly have socially focused missions, such as contributing to civic society[12] or to regional economic growth.

The helpful thing about mission-driven approaches to impact is that they typically give you the endpoint, and often clear measures of change, from which to work backwards. SDGs are very clear on what 'global better' looks like, charities are very clear on the vision of 'better' for their community, and organisations have a clear remit for which they might find research helpful. The job of research in each instance is to help get there.

Personal Motivation for Impact

For many people there is a far more fundamental reason to do impact; a personal wish or sense of responsibility to make things better. Personal motivations can be characterised by the question *how can I make a difference?* This might arise from personal experience, vicarious experiences from those around us, or simply being allergic to injustice and inequity. Many of us have gravitated towards impact for this reason, driven by an intrinsic need to contribute in whatever way we can to solving problems and making

---

[12] For more on 'Civic Universities' see https://upp-foundation.org/about-us/civic-university-network/, and an example of university commitment here https://www.lincoln.ac.uk/media/responsive2017/documents/the-new-civic-university-university-of-lincoln.pdf

a difference. That might be idealist, but I'm yet to be persuaded against the awesomeness of the collective power of trying.

Perhaps the simplest way to categorise these drivers is:

We must: external and instrumentalised requirements such as funding or assessment which require or even mandate impact from research, whatever this impact may be. Arguably this is the driver which varies most by location, as the impact agenda looks and feels different across nations.

We should: stemming from broader missions, such as SDGs and university's Civic agendas, which imbue a sense that it is incumbent on us to contribute to society, using the engine room of the university to drive change.

We/I want to: personal motivation, often relating to passion around the subject area, lived experience, appetite for social justice or other such drive to contribute to change.

An individual could of course be influenced by several of these drivers at the same time – personally motivated to deliver on a broader mission whilst also needing to apply for funding/create a case study. Where these align the effect can be supercharged, but where they don't it can bring tensions.

THE WONDERFUL WORLD OF IMPACT TERMINOLOGY

Terminology can be both helpful and inhibitive when it comes to impact. 'Impact' is often used as an umbrella term for *any* connection between university and the 'real world', sometimes used interchangeably with engagement and knowledge exchange. But as we've already defined, impact is the change not the process. Obviously no one owns the word impact, nor is it used exclusively for the research–society link, but an absence of definitional boundaries in relation to research impact risks everything becoming too blurred to mean anything at all. Impact is, as far as I can see, one of the only words in the formal academic lexicon that means benefits to *real people*. Knowledge exchange, dissemination and engagement (and many others) all relate to the actions WE take, in pursuit of – but not necessarily extending to – making an actual difference. Without language which locates our actions outside of the academic bubble, we risk becoming self-congratulatory and disconnected from the wider world.

> *'At first glance you might think 'what's in a name?' and that spending time on definitions in conversations with researchers is a waste of time. However, also having this discussion increases impact literacy. It helps you approach impact more strategically (yes, I know that sounds neoliberal but bear with me). Intrinsic motivation for impact is all well and good but not everyone finds that nugget of gold so easily. Training researchers in the art of dissecting the funder's or stakeholder's language provides more clarity, might increase efficiency in planning and writing. Also, from the point of view of a university developing policy and support on impact, using terminology in a consistent way helps the common understanding and uptake.'*
>
> Esther De Smet, Senior Policy Advisor, Research
> Department of Ghent University

Impact is also used as both a singular and plural term. Impact is commonly used as a capture-all for the broad agenda of research-led change, but once we get into the detail, we see research actually has multiple impacts in the real world. In this book – as with much of the general usage in the sector – I'll use impact (singular) to reflect the broad topic, and impacts (plural) to reflect the range of societal effects we might see from research.

To my knowledge there is also no *absolute* set-in-stone internationally accepted impact glossary as terminology is used variably across the sector. There are of course guiding definitions offered by various funders, government departments and others, but it would be unrealistic to say they were fully harmonised. There seems to have been a terminological shift over the years in recognition of that the act of impact is not simply a provision (transfer), or transaction (exchange) between research and society, but a more active and negotiated affair needing active *knowledge mobilisation*. And in the UK at least, the term 'knowledge exchange' has become more synonymous with the breadth of ways in which a university contributes to wider society[13] rather than activities in service of research translation alone.

Within this, the dominant use of *impact* is as a description of the actual effects felt outside of academia. And it's vital we keep this as our line of sight to avoid us falling back into definitions which leave us circling within the walls of academia. My mum really isn't going to care, other than in a gloriously 'well done' mum tone, that I've had a paper published or my *h*-index has gone up. My friends aren't going to care if I tell them how far up whatever ranking-is-in-favour we've improved in as an institution. These things have credit within academia, but are arguably not really meaningful to any extent elsewhere. People care that our research is making a difference – new drugs, social justice, lower emissions (carbon not personal), more job opportunities, businesses retaining staff, etc. It's easy for us to sometimes forget with – academia being such a pressure cooker of introspection – that the things we count within the university walls often don't count outside. None of this is to suggest the things we do within academia aren't without merit, just that from a societal perspective they're far from the main focus.

I'd summarise the situation as saying that whilst we have some clear vocabulary about impact, it is not systematically or consistently used. And that's a problem because it makes it far harder for people to get a handle on it, and can create those awkward tensions between people as they realise too late they've been saying the same word but meaning a different thing. Clear vocabulary can help us not only understand but connect with others about impact, yet we can get so unhelpfully tied up in precision that we immerse ourselves in semantic debate rather than just getting on with it. So for that reason I'm not going to take us into a labyrinth of definitions, but invite you dive into that world as you advance in impact, exploring the work of far smarter academics than me. Instead let me offer this set of broad, guiding definitions to give you a mental bookmark for the various and interrelated concepts:

- *Knowledge production*: The act of generating new knowledge, that is, research. This is synonymous with terms such as explo-

---

[13] Knowledge Exchange Framework website https://kef.ac.uk/

ration, investigation, experimenting, observing, immersing or other disciplinary style words to describe research.

- *Research process*: The activities and efforts which take place to deliver the research.

- *Outputs*: The things that are produced by the research, such as journal articles, performances, interventions, tools, methods, exhibitions, widgets, etc.

- *Outcomes*[14]: The new knowledge 'end results' of our research, for example, what we now know/have confirmed/have validated/have unveiled/what we now have available for use/what has been discredited.

- *Dissemination*: The act of communicating, broadcasting or sharing research outcomes.

- *Knowledge mobilisation*[15]: The act of *actively* connecting with and engaging society at any point in the research process. Related terms include knowledge exchange, transfer, engagement, brokerage and outreach, each with different connotations about how active this link is, and all essentially more than 'just shouting about the research'

- *Engagement*[16]: Active connection with non-academics to shape, deliver, share or use research. Types include:

  o *Public engagement*: Actively connecting with the public.

  o *Policy engagement*: Actively connecting with policy-makers or the policy process.

  o *Business or industrial engagement*: Actively connecting with businesses.

  o *Charity engagement (you can take it from here....)*

---

[14] There are varying views on how outcomes and impacts overlap, with some treating impacts and outcomes-embedded-over-time, and others (like me) viewing outcomes as the 'thing we get from' research, *which enable* impact.

[15] You will see knowledge mobilisation, engagement and dissemination overlap. Don't get bogged down in the terms, just focus on actively and meaningfully engaging

[16] Engagement is listed specifically here as it's a commonly used term; in practice knowledge mobilisation needs engagement, and engagement is a form of knowledge mobilisation

- *Coproduction or cocreation*: The act of working *in partnership with* non-academics to conceive, design, deliver, analyse, interpret, disseminate and/or implement the research.

- *Impact*: The changes or provable benefits of research outside of academia.

- *Indicators*: Things you can point to show ('indicate') impact has happened (e.g. *more customers*).

- *Research evaluation*: Assessments of the inputs, processes, outputs and outcomes of research and those who perform and enable it, for the purposes of (e.g.) analysis, accountability, advocacy, acclaim, adaptation, allocation, recruitment or promotion.[17]

- *Stakeholder*[18]: Anyone who has a particular interest in the work, be that direct or from a distance.

- *Users*: Stakeholders who use the research in some way.

- *Beneficiary*: Stakeholders who benefit from the research in some way. *NB*: Sometimes you might find words like 'direct' or 'ultimate' attached to the term beneficiary (e.g. in funding guidance), respectively, relating to those who most immediately benefit from the research, or those who are the overall target for any impact.

Now I'm very aware that I'll be opening a can of worms with any strict definitional boundaries. The definitions above are intentional simplifications to help you understand the scope or spirit of different bits of the process rather than suggesting they are absolutes. The trick is to recognise what each is, and then consider how they knit together in your world.

> '*If impact is the destination, Knowledge Exchange is the car. The research gets in the car, and it's driven to somewhere it could make a difference. But that's not enough; it can only make that difference if it actually gets out of the car and does something. If you drove Route 66 and never got out of the*

---

[17] Huge thanks to the guru Dr Elizabeth Gadd for this definition.

[18] There are some reservations about the term 'stakeholder' due to a historic relationship with colonialism. At this point, there isn't a clear alternative term which covers the same ground, so I'll leave stakeholder in place for now ahead of wider debates.

car, you could only say you drove past the World's Biggest Ball of Yarn, but didn't actually see it. And if you don't have the ticket stub – and if KE activities aren't followed up and evaluated – how could you prove it?'

Helen Lau, Associate Director for Knowledge Exchange, Coventry University

'[...] taking this analogy further, funding is (commonly) the fuel in the car. Using a travel agent to plan the journey is like using an impact officer to plan the impact strategy. Calling ahead to find out what attractions are open is like stakeholder engagement. And if a local person offers to take you on the tour, that's co-production.'

Dr David Phipps, Assistant VP Research Strategy & Impact, York University (Toronto), and Director of Research Impact Canada

Things That Sound Like Research Impact But Aren't

One of the reasons definitional clarity is so helpful is that without it we end up with misinterpretation and misunderstanding. To that end there are three main places I find there are things that sound like impact but actually aren't:

1. *In the term 'Academic impact'*: This is a not uncommonly used, particularly within funders, to describe the influence of research *within* academia on things like disciplinary knowledge, methods, paradigm shifts, new models and new theories. Influence *on* academia itself is of course hugely important, and it's not uncommon for some agencies to share the word impact across these two areas to semantically equalise the value of real world versus scholarly contribution. The problem is that it creates the opportunity to conflate academic prowess with benefits to society, at worst allowing us to congratulate ourselves without making any changes in the world at all.

2. *Impact rankings*: There are a range of ways in which university performance is measured, with an increasing interest on their contribution to the wider world. These are commonly related to

SDGs, with institutions scored on various criteria to create met-ricised rankings of their contributions to the 17 areas of change. This of course resonates with and in places crosses over with impact, but caution is needed to not conflate wider institutional contribution with benefits which arise specifically from research.

3. *'Journal impact factor (JIF)'*[19] *or just 'impact factor'.* JIF have been a longstanding feature of academia, offering a short-hand calculation of the relative influence of a journal. The JIF reflects how frequently on average articles in that journal have been cited in a particular year and is calculated by dividing the number of citations in a certain year by the total number of articles published in the two previous years. The higher the number, the higher the average number of citations. For many years this number sat as a proxy for the quality of the jour-nal, and was commonly used by academics to select journals, by institutions as part of individual or research assessments. In recent years, there has been an increasing resistance to the blunt dominance of JIF, particularly given it cannot indicate the influence of a specific article. Whilst there have been a range of more nuanced bibliometric alternatives developed as more focused markers of influence, from an impact perspective they still relate to academic attention, rather than real-world effects.

Bibliometrics feature strongly in all three of these, and within academia, bibliometrics are still big business. Whilst their weight varies by discipline, country and other factors, measures such as citations, *h*-indices and JIF are being increasingly recognised as inappropriate markers for impact. There is perhaps nothing inherently wrong with metrics relating to publications, insomuch as they reflect attention within the academic community and act as a marker of the creeping accumulation of knowledge through the research evidence system. However, there is an increasingly strong push back on the *power* of metrics[20] within academia,

---

[19] For more information on the calculation see https://www.nihlibrary.nih.gov/about-us/faqs/what-are-journal-impact-factors

[20] Check out the incredibly important work on Responsible Metrics here https://responsiblemetrics.org/

particularly in relation to their irresponsible use. Even stronger, arguments relate to metrics being weaponised as a means to judge academics (within progression and recruitment) and institutions (in rankings), dismissive of both the game playing and power such approaches have. The nuanced arguments and political debates about the comparative merits (and damage) of metrics-led approaches are far beyond the scope of this book, and beyond impact, but I let me nail my views to the post and say that measuring academic(s) value by publication numbers is in my view foolhardy, reductionist and a big can of stupidity wrapped up with a stupid bow.

The issue for impact is that bibliometrics are a commonly confused crossover point when talking about impact measures. I say crossover slightly ironically as there isn't really a crossover at all. One of them measures academic attention, through pre-set algorithms, where the unit of measurement is the academic publication, and effort is reducible to single and comparative numbers. The other is impact. Academic referencing of an output might mark a step along the route to impact, but it can only demonstrate scholarly attention for an output, *not* real-world change. Compare that to impact measures, which are anything which show change beyond academia. These might be quantitative or qualitative – whichever is more appropriate – and might depend on output, or have nothing whatsoever to do with a neatly formatted pdf. Outputs are extremely important within the impact landscape, as 'vehicles' of the new knowledge which can be mobilised, and as substantiations that the research was undertaken, and we all know they're part of what we do. But unless we're talking about citations in something outside of academia (such as policy), we're confusing academic attention with real-world change.

I'm still saddened by how many people shout 'citation' in response to questions about measuring impact. It's perhaps unsurprising given the foothold of metrics in the measurement of academic prowess, but nonetheless the Citation God Complex is still dominant enough to cause many of us to quietly (or loudly) sob. I remember once having an argument with a publishing exhibitor at a UK research management conference:

- *Exhibitor*: 'Impact? Oh it's citations'.

- *Me*: 'Ah no, impact is benefit in the real world'.

- *Exhibitor*: 'No, it's citations'.

- *Me*: (with a mild scowl) 'No, that's attention. Impact is when real people or things benefit, and citations aren't that'.

- *Exhibitor*: (looking at me condescendingly) 'No it's definitely citations'.

- *Me*: (now fully scowling) 'This is a research management conference, full of impact people, with a special interest group on impact. Impact in this context is very definitely "real-world change", not citations'.

- *Exhibitor*: 'We'll have to agree to disagree'.

Feeling the futility of continuing, at that point I did what any self-respecting impact person would do – swept as many of their conference freebies as I could into a bag and walked off. I have also passively aggressively refused to tweet about anything I've ever seen them do at a conference since and have routinely tried but failed to send colleagues to acquire their freebies for me at subsequent events (I still want free stuff, obviously). Clearly I'm not declaring my judgement should never be in question – I'm not David Attenborough – but the point is that even when told, *from within the community*, that the definition of impact was wrong *for that community*, they still refused to reconsider. Posturing instead the issue must be my poor understanding, compromising only so far as 'agree to disagree'. Sadly obstinate views like this just continue to de-centre impact narratives away from what matters to real people and annoy the hell out of most impact folk as they diligently acquire free pens.[21]

> '*One of the problems for impact is that so much of it is driven by agendas that are monitoring and auditing based, not impact. Funders want to report on a periodic basis; often*

---

[21] Incidentally I've never used the notebook I *cough* acquired. Partly because of my annoyance with the company, and partly because even more annoyingly it's really nice and I'm saving it for best.

this means by an annual publication to key stakeholders. Busy people ask other busy people for information for that report before an immovable deadline is hit. Collecting and counting numbers is easy compared to the complexity of generating a meaningful impact story so impact can be driven by the annual reporting cycle and default to counting things. Impact then begins to resemble a game of Numberwang, that fictional TV series where two contestants shout out seemingly random numbers which are occasionally told to be 'Numberwang'. Simply put the research funding's ecosystem can promote an unconscious bias towards the cult of counting, confusing audit for impact and missing the golden rule; impact is best described, with numbers that help underline the story, but not instead of the story itself. We all understand this, but the day job and cultural norms get in the way.'

Dr Mark Taylor, multiple sclerosis patient advocate

DIMENSIONS OF IMPACT

For a deeper understanding of impact, we need to recognise it's not a single big thing, and instead explore its dimensions. As elsewhere in this book, it's worth saying at the outset that there are (to my knowledge) no set-in-stone definitions of many of these things either. It is entirely possible – and fairly enjoyable in you're so minded – to go down a definitional rabbit warren on these terms, particularly as in practice they are operationalised in varying ways and thus open to semantic debate. But for now here's some simple and differentiating definitions to help get a mental foothold on how impact works and how it's judged.

Significance: How Important Is It to the Outside World?

Within both assessment and funding, impact is typically measured in part on its significance, or how *important* it is. This is of course subjective, but let's park that and consider how you could determine and articulate what significance looks like. Crucially significance here relates to how important the issue is *in the real world* rather than academic ideation. Therefore, the substantiation of significance needs to come from that same world. This doesn't mean that because we're in academia we can't know if it's significant, purely that claims of importance must stand up to scrutiny to those outside the research bubble. Common ways to demonstrate significance include referencing relevant things like policies, strategies, organisational reports, data and statistics, research findings (where they show something is needed) or find out first hand from the people to whom it matters.

Reach: How Far or Deep Is the Effect?

Another aspect of impact is its reach, which is *how far or deep the impact is*. It can be easy to assume this is about numbers (marked by metrics which show 'big') and geography (marked by international effects) but societal change is far more nuanced than that. Sometimes size does matter – 100 patients treated is better than 50, and a business will definitely prefer to have £1 million rather than £100 profit – and we should never not be seeking to make the world a better place. But reach can never be taken out of context. Big numbers are only a marker of reach when the impact itself is about size, and geography is only a marker of reach when the impact particularly matters 'there'. What about if the impact relates to something more subjective? Or if numbers can't really express the depth of a change? Or if the change we need to make is very close to home? Instead of thinking of reach in terms of numbers or geographies, I'd recommend you think in four dimensions: horizontal, vertical, depth and over time.

- *Horizontal reach: Spread and proportion*: Sometimes reach can be defined by how far the effect goes. For example, numbers of people trained or numbers of schools using your new teaching resource. But assessments of reach also need to reflect what could realistically be expected in a particular context; if you're working in rare conditions, there's no way you could hit the same 'numbers' as for an illness affecting millions of people. Sometimes it's not just about the reach itself, but what this represents *in proportion* to the size of the context it's in. Consider: *how far does my research get in context?*

- *Vertical reach: multi-level*: Sometimes instead of a wide (horizonal) effect, research can have an effect at different 'heights' of an issue. Analogous to 'from factory floor to boardroom', reach of this type is through the combined effects of ground level practices (such as staff training) through to strategy or policy change. Consider: *does my research address the problem by changing things at multiple levels?*

- *Reach as depth*: Sometimes the impact focus of research is not to go big, or high, but to go *deep*. Typically this kind of impact dives into areas often left unaddressed, such as amplifying the voices of those rarely heard, increase the visibility of those rarely seen, or otherwise lifting an issue from the conceptual seabed. Depth requires a clear sense of context and social significance and is characterised less by spread or levels of change and more by the extent to which something missing is now seen. Consider: *does my research bring something to the surface?*

- *Reach over time (unlocking a cascade)*: Rather than create big effects directly, sometimes research injects a change which unlocks the potential for effects over time. For example, if your research prompts a change in a definition within the legal system, this could enable laws to be applied in a different way or extend the circumstances of its application. The research opens the door for numerous onward changes – *a cascade* – with reach determined by how much difference is made by the initial 'unlock'. *Consider: does my research open a door to change for others?*

A single project can achieve reach through any and all of these paths, and reach can be increased (scaled) by following any path along its logical trajectory – doing more of the same (horizontal), addressing different levels of a problem (vertical), diving further (depth) or following the onward benefits of anything you've unlocked (follow the cascade). This list isn't and can't be exhaustive, but should give you a sense that reach is far less intrinsically about numbers and location, and far more about increasing meaningful change in whatever way that makes most sense.

Contribution and Attribution: How Much of the Change is Down to the Research?

I've grouped these here as together they reflect the way in which research contributes to, but may not be the only influence on, societal change. There is no doubt research offers a VERY important contribution to society, but it would be unrealistic to suggest it's always the only active ingredient in social change. Sometimes research is the main protagonist, but sometimes it's the equivalent of a character who sweeps in during Act 2 to reveal a pivotal plot point which unlocks the mystery. Our challenge therefore is gauging how much of any effect is due to the research, and how much would have happened anyway. To reassure you, this isn't about locking in a metric or percentage, but about making a reasonable claim (i.e., avoiding under-or-overclaiming the impact). We do this by considering two related things:

- *Contribution*: What research actively 'adds to the mix' – *can be shown to have contributed to* – societal change.

- *Attribution*: The extent to which the impact can reasonably be claimed to have arisen from – *can be attributed to* – this contribution.

In the neatest scenario, the research is the only active ingredient in driving change, meaning impacts can be fully attributed to the research, that is, the impact could *not be possible* without the research. For example, if you create a new drug for condition X, and there was no such treatment for X before, then the treatment benefits must be attributable to the new drug. More commonly however, research is one of the several influences on change. For example, if you design a new fitness intervention for school children, this might be added to the curriculum alongside activities on healthy eating and getting a good night's sleep. If the children are found to be fitter and healthier, how much of that is down to your research? The blunt answer is you can probably never know, but by working with the teachers you can explore what your research added, and gauge what would be different if your research hadn't played a part.

Contribution and attribution overlap, and you need both parts for a full impact claim. Attribution can only be possible if the research has contributed because – radically – you can't determine how beneficial something's been if it didn't do anything. And the extent you need to prove each may vary by the agenda. For research assessment it is typically necessary, but not sufficient, to demonstrate contribution, needing also to substantiate claims of attribution and provide evidence of the effect itself. For funding bids, narrative descriptions of the way in which research is planned to contribute to an impact may be sufficient. And the longer an impact journey takes, the more attention you might need to pay to tracking the way your research progressively contributes, and/or forms part of a larger set of actions towards change.

Distance and Time: Where and When Does the Impact Happen?

I've grouped these together here as both terms reflect the length of the golden thread of impact, be that in time or geography. There are very few rules about where and when impact must happen, except that:

- The 'when' cannot predate the research commencing; and

- The 'where' must be outside of academia.

With regard to 'when', impact can happen during, immediately after or much later than the research. The only fundamental rule is that it can't predate the research, as unless you're Doctor Who and have time travel capabilities it would be impossible for impact to come from research that hasn't happened yet. But there's no rules that it can't happen alongside – *be concurrent with* – the research. If you think about participatory research or research in collaboration with external partners, the act of being involved can itself be beneficial. For example, young people involved in arts projects could develop skills (capacity building) or construction businesses involved in environmental projects may change their mind about the importance of sustainable materials (conceptual change). In contrast, impact may not occur for many years, possibly because it has a long and scripted path (such as with drug development), the world is just in some way not ready for it yet (the 40-year Higgs boson delay anyone?), or because a brilliant idea under one political party is considered a dangerous move by another (*insert your own political annoyance here*). And it can also be because amidst the pressures of everything else, we take our foot off the impact accelerator to firefight other things like annual appraisal deadlines and Reviewer 2.

With regard to 'where', whilst we can casually talk about impact needing to be felt outside academia, we can't overlook the different lengths of journeys for different types of research. Some research is inherently much closer to use (e.g., applied research), some has a longer race to run (e.g., fundamental research), and some is already waving from the sunny side of the impact gates (e.g., participatory research). When it comes to your starting location, *it is what it is*, and whether you're closer to or further from the impact zone is just an artefact of the kind of work you do. And when it comes to your destinations, you are joyously unconstrained by location (local through to international), whether it's in the same domain (e.g. social research having a social impact) or if the effects cross into a different domain altogether (e.g. social research having an environmental impact). Mix and match as the world needs. Start where you need to and end where it matters.

Linearity and Dependencies: How Sequenced Does it Have to Be?

Sometimes the rhetoric of impact can suggest that all pathways are linear. Actually other than the necessary linearity of time (see Doctor Who above), there's no rules about how linear or – and with apologies for a technical term here – wibbly wobbly the impact paths for any particular research project must be.

Much of the default expectation for linearity seems to come from the historic roots of impact in knowledge transfer and 'return on investment' economics. These can convey a sense that there is an overarching sequence for translating research from concept through to large-scale effects, with impact linearly calculable at the end. Many impact frameworks which outline a single straight route from idea to wide scale effect are great for giving us an aerial snapshot of a generic route between research and impact. However, as soon as we try to apply them prospectively, we can end up trying to shoehorn our research into a process which doesn't quite suit the context. This problem is magnified in research yielding multiple types of impacts, in domains other than the core subject area, or with impacts jumping out of the research process itself. If we only count what happens at the end, just like the Hitchhikers Guide to the Galaxy problem, we might know the answer is 42, but do we even know what that means?

Accepting some paths are far less 'as the crow flies' than others, it's more useful then to shift from default assumptions that paths are linear, to spotting those which can *only* be linear. Some research has pathways to impact with inbuilt sequences and thresholds which must be passed for impact to happen. Some of the clearest examples of this are:

- Research which needs to demonstrate safety or efficacy to an acceptable standard (e.g. through a clinical trial and laboratory testing or simulation).

- Research which needs to produce something to a pre-defined level of quality (e.g. a specific level of engine efficiency).

- Pathways which need regulatory approval, legal protection the right political winds.

If your research fits into this kind of category, then your route to impact already has a script. But if it doesn't, enjoy the opportunity to follow paths in all different directions.

DISCIPLINARY DIFFERENCES? *NOT AS SUCH*

Research is incredibly varied, in type, purpose, translational potential, methodology, topic, sensitivity and all else. By definition, the routes to impact will also vary, often with more protracted pathways for research at the fundamental (vs. applied) end of the research spectrum. But thinking about the differences in impact paths as differences in disciplines *per se* can mask similarities between fields and variations within subject areas. Instead, therefore, it can be more useful to think about variation by how close to (or far from) use the research is, how integrated (or not) non-academics are in the process of research, and how accepted (vs. contested) the impact goals are. Trying to outline all the possible permutations of this would be eye-achingly hard, so instead below are some illustrative 'profiles' of different types of research to give a sense of how pathways not uncommonly play out.

Fundamental or 'Discovery' Research

This is research which tries to discover a piece of knowledge (sometimes called fundamental, exploratory, discovery or 'basic' research) and is typically characterised by activities to unveil or identify a block of knowledge. It is often at the start of what some would call the translational chain of research and is commonly (but not exclusively) of most immediate interest to others in the academic community. Fundamental research commonly needs an output (e.g. journal article) to demonstrate its rigour and communicate the findings if it is to be used by others. If your research falls into this category, ask yourself:

- Could I do follow up research which helps it head towards an 'application'?

- Are there other disciplines, or more applied versions of my own discipline, I could pass this to?

- Can I draw in the skills of others to find ways to translate it into use?

Philosophical Research

This is research at the more philosophical, epistemological or ontological end of the scale (whether that's within the discipline of Philosophy or not) which seeks to raise questions or query assumptions rather than identifying 'cold hard facts'. Its route to impact is often by supporting changes in thinking (conceptual impact) in others, helping them to question or reshape a perspective on a topic. If your research falls into this category, ask yourself:

- *Who* could make use of this new way of thinking?

- How could you *engage* people with this new way of thinking and make it usable?

- What 'cascade' would these conceptual changes unlock?

Participatory or Engaged Research

This is research in which a study is dependent upon the *active* involvement of those outside of academia, be that through participatory engagement methods, creative practices or any other means. Participatory methods exist across many disciplines, but are perhaps more common in arts and social sciences where more qualitative approaches abound. As with all research, engaged methods can have a long-term impact legacy, but this type of research is particular marked by the ability to create immediate benefits to those involved through the process itself (skills, knowledge, networks and so forth). If your research falls into this category, ask yourself:

- What are the direct benefits for those involved?

- How can I capture the immediate or direct benefits to those involved?

- How does this engagement enable further impacts?

Research Which Aims to Develop a Useful 'Thing'

This is research where the goal is to create a specific useful 'thing', such as a prototype, product, component, intervention, manual, guide or framework. It can be arrived at by any method, but characteristically results in something with a clearly defined purpose or use. If the useful thing is designed for use outside of academia – such as a new diagnostic tool – impact starts as soon as it starts being used. However if the useful thing is designed for academia - such as a new research method - then the impact path depends on it (a) being used in further research, and (b) this enabling impact. If your research falls into this category, ask yourself:

- If my 'users' are academics, how can I follow their use of it and the onward benefit to society?

- If my 'users' are non-academics, how can I follow their use of it and how they benefit?

- Have I considered what existing 'things' would need to be taken out of service (*de-implemented*) to make room for mine?

Research in Contested, Sensitive, Taboo or Secret Areas

Where the topic is, for some reason, sensitive, this can influence not only whether you can communicate about it, but how much resistance you might hit. Research can be sensitive for a number of reasons, such as commercial, political or security concerns, or because the topic is controversial, ethically unclear, stigmatised or otherwise contentious. Where it can be communicated, there is likely to be pushback, dissent or disbelief, and where it can't be communicated you may have to rely on single pathways to impact (such as use by the military department who have commissioned the work). There are typically protocols in place within assessment to accommodate research which can't be disclosed, but it can still be challenging to determine how to get traction on something not everyone can know about or agrees with. If your research falls into this category, ask yourself:

- What can or can't I communicate, and what routes to impact do I then have?

- Am I seeking to have an impact *on* the thing that's sensitive (e.g. change the way it's thought about) or do I need to work *around* sensitivities in trying to implement it?

- What measures could I put in place to deal with, and be supported in dealing with negative reactions?

Commissioned Research

It is not uncommon for researchers to be commissioned to study something for an external organisation. Such research typically has a fairly defined scope (why would they commission it otherwise?), a clear rationale, and a sense of how/why it could be used. Unsurprisingly commissioned research can be a brilliant route to direct impact, both via direct benefits to the commissioner and indirect benefits (how they then go on to use it). The downside is that you might be locked into the organisation as the 'only' user. Imagine that you are commissioned to develop a more efficient car battery. You do your clever thing, and the delighted company take it to their Board who decide they will fully transition to your new battery in all of their new build cars within three years. *Boom* – impact on their strategic decisions, on the car design, and ultimately on any sales and environmental benefits the new battery enables. But the company don't want their competitors to know about it so ban you from publishing. And then the company goes out of business. What was a lovely strong route to impact has now become an unrealisable promise. This is worst case scenario of course, but commissioned research jointly brings the opportunity for direct application and the risk of 'failure to thrive'. Thankfully much of this can be managed contractually and with intellectual property expertise, working with the experts within your institution from the outset. If your research falls into this category ask yourself:

- What is the route to impact through this commission and how can I make it most likely to happen?

- What can or can't I publish (i.e. how open is my impact route)?

- What measures can I put in place – with help – to manage the risks of putting 'all my eggs in one basket'?

Research to Curate, Preserve or Order Knowledge

Research of this type seeks to gather together, synthesise, coordinate or consolidate knowledge. It is characterised by activities which seek to find the pieces of information by which we can 'know'. Common examples include finding, translating or preserving historical documents, or reviewing existing research knowledge to answer a new question (e.g. systematic reviews). Its route to impact is related to the question *who can make use of this knowledge?* If your research falls into this category, ask yourself:

- Why is this curated knowledge needed?

- What does this coordinated knowledge enable that wasn't possible before?

- Who could make use of it?

These research 'profiles' are offered to help you think about how the shape of your research – not just the topic – can have a particular kind of relationship with the wider world. Research is too joyously eclectic to cover in a few short illustrations, but by at least trying to picture how you and others do research, you can also think about ways to connect skills, share opportunities and think outside the proverbial box.

> '*Whatever your discipline, speak to colleagues in other subject areas. Creativity and innovation is rife in academia so embrace it and talk to folks who aren't your usual pals.*'
> Lorna Wilson, Co-Director of Research & Innovation
> Services, Durham University

WHAT COUNTS AS 'BETTER' IMPACT? (IF YOU NEED TO PICK)

Before we close this chapter I want to take a moment to talk about how, amidst this brilliantly broad church of possibilities, it isn't at all uncommon (particularly in assessment-driven or resource limited settings) to need to narrow down your potential impacts to the *best ones*.

When considering what counts as 'better', it won't surprise you that I'll pre-emptively caveat this section with *it depends on who's asking*. 'Better' could relate to the score received by a case study, the fundability of an impact plan, or how significant those in society judge the changes to be.

Better, in its truest sense, *always* stems from what matters to those outside of academia. But for many of us, we also need to consider how to articulate 'better' in more instrumental terms. Thankfully, it's here that understanding the dimensions can help, and being able to calibrate 'how impact works' for your type of research can both help good practice and also dissuade you or others from pushing for impact that can't reasonably happen. Whether you need to identify impact priorities in bids, for wider missions, or in assessment, accounts of 'better' impacts tend to be the ones that:

- Address a clear and important need (show *significance*).

- Generate a wide or deep effect (show *reach*).

- Have a clear and compelling link between research and effect (show *contribution and attribution*).

- Articulate clear changes (show *impacts*).

- Use/indicate appropriate measures and evidence (show or point at *proof*).

- Identify and engage people and organisations (engage *stakeholders*).

- Actively rather than passively connect research with society (show *knowledge mobilisation*).

- Have clear and realistic pathways, including recognition of both opportunities/facilitators and barriers.

Unsurprisingly, lower rated accounts typically exhibit the opposite, that is:

- Unclear or unsubstantiated need, relying on the reader to accept the 'problem' and the assertion that the research solves/d this.

- Small or narrow reach, where this could reasonably be wider.

- Disconnected, unjustified, unrealistic or assumptive paths.

- Forge no, or only a weak link between research and impacts.

- List unclear, unachievable, unrealistic or meaningless impacts/goals.

- Passive over-reliance on research quality, dissemination or academic reputation being 'enough' for impact to happen.

- Little or no connection with stakeholders.

- Scattergun approaches to indicators, evidence and impacts.

Impact is very often reviewed by both academics and non-academics, so for both plans and case studies your goal is to assemble a narrative (and evidence where needed) which leaves them in no doubt about how important the impact is, and how it can be/was achieved. When it comes to conveying the 'best' impact, remember that however crystal clear the significance is to you, and however passionately you feel about it, unless someone else can be assured of this you'll struggle to make any impact case.

SUMMARY

This chapter has done the heavy lifting of explaining what impact is (and isn't), and is perhaps the most textbook like part of the whole book. So what have we learnt?

Impact is the provable benefits of research in the real world. It's the word for the change research brings about, rather than the actions we might take to try and make it happen. Whilst universities contribute to society in a lot of different ways, impact is specifically the conversion of research into benefit outside of the academic bubble. Impacts can be changes in a thing (instrumental), a way of thinking (conceptual) or the ability to do something (capacity), can happen in any 'domain', and can arise from research findings or the process itself.

Assessment is a dominant driver of impact where it is in place, with funders' requirements giving similar impetus for plans and efforts in pursuit of social change too. More broadly, societal agendas – such as SDGs – can set a clear line of sight for the importance of collective action, but amidst all this personal motivation remains a powerful driver of why we push onwards in trying to make a difference. The precise rules and governing structures for impact will vary over time, by geography, through political change and any way in which the system chooses to reconfigure itself, but impact remains the expression for how we make a difference through research, whatever bells and whistles are added after.

Different types of research may have different routes to impact, but variations aren't inherently bound by disciplines. Research where non-academics are involved in its delivery has the opportunity for immediate impacts, whilst other research must wait until it has 'final results' to share. Some research has scripted and linear paths to impact whilst some can map its own scenic route. Research at the applied end of the spectrum can more immediately be translated into use but that at the fundamental end offers the building blocks of knowledge researchers may use as the foundations for impact paths. Research which addresses a significant need, and can be openly talked about with an identified audience has a clearer path to impact than controversial research or that trying to drive change society isn't yet ready for. And what counts as 'better' is a matter of perspective.

By recognising that impact is not just a thing done for funders or assessors, and that the journey to impact can be all kinds of shapes and sizes, we can really start to recognise the opportunity we have for designing our own pathway towards change that matters.

WHAT CAN YOU DO?

With so many different drivers, terms and dimensions, it's not surprising impact can feel like a big unfathomable blob at times. Taking time to understand what impact is (and isn't) is one of the most powerful steps we can take to help ourselves. Take a step back and look at what's motivating you and the institution you work in. Think about what your research offers to the world and how you can express its significance and contribution. Visualise the route to impact, be that a straight, jagged, short or long line. Lock into terminology that helps, and hold onto the core point of impact – provable benefit of research in the real world – as you try to connect the puzzle pieces of academia into a picture that makes sense.

It's worth it.

Ask yourself:

- What motivates me to do impact?

- Are there formal requirements for impact at a local, national or international level?

- What are the conceptual, capacity or instrumental changes my research might lead to?

- Does impact arise from my research findings, the process or both?

- How can I substantiate the impact need for my research?

- Where and how does my research 'reach'?

- Is my route to impact linear?

- When are impacts likely to start happening?

Chapter 2

IMPACT LITERACY

Impact is essentially the lovechild of research and need.

It is perhaps odd that a book on impact literacy waits until Chapter 2 to get to its main plot, but it's important to have established some of the basics of impact before we can launch into what being impact literate means. I called this book 'developing an impact literate mindset', because I genuinely believe that doing so is the starting point for doing impact well. But what does that mean?

Let's start by explaining how this work came about. Picture the scene. It's a cold October day in Edinburgh, Scotland. The glorious David Phipps and I had been corresponding on Twitter (*is 'corresponding' the right word for twitter chats? DMing? My kids remind me I'm not cool*), and unknowingly found ourselves at the same event for impact officers in the UK sometime shortly after the REF 2014 submission. We sat, almost transfixed at the level of despondency, angst and bereftness of colleagues within the UK impact community, utterly worn out from the labours of submitting case studies to the most recent assessment cycle. Colleagues talked in such negative terms, some even declaring they would leave their jobs, a shared sense of exhaustion and unacknowledged damage.

As the morning progressed, we both – separately at that point – felt the despair so intensely that our first words to each other, in sync at the lunch break were *this has to change.* That microcosm of people had shone the strongest of lights on how industrialised impact had become in parts of academia, a far cry from the simple rhetoric of simply 'showing what difference research has made'. An army of people working not fully on *impact*, but on devising time- and institution-bound rule-compliant accounts of world-changing effect.

Perhaps more profoundly, it foregrounded for us that the pressure to generate accounts of research-led impacts concealed the scale of effort needed to broker research into practice. The result? Impact reduced to stories of assessable effect (rather than of meaning); effects attributed neatly to research (as though research had its own lifeforce), and energies disproportionately expended agonising over how to curate, persuade, impress, filter and corroborate an invariably longer and messier process than could be contained in final condensed print version.

We were lucky enough that David obtained a fellowship from the Association of Commonwealth Universities (ACU) for a study visit to the UK. If you know David, then you'll know his energy is utterly infectious, and I'm still astounded to this day that our mutual jazz hand'ing hasn't yet triggered some kind of typhoon. Anyway, he and I spent the visit trying to see what we could do to help in any way ease the tensions, exploring potential ways to introduce better and healthier approaches to impact to offset not only the pressures on an already tired sector, but also the ways in which impact rhetoric had deviated from a meaningful focus on real need.

That first visit was absurdly helpful – I am still so grateful to have had that space and time at that point in the evolution of impact within the sector. We reflected on the differences between the UK and Canadian research systems to see if we could glean any possible routes through and noticed a particular quirk. In Canada, emphasis was put on the commitment to engagement, weaved into expectations by funders with an absence of a formal impact assessment agenda. Conversely in the UK, emphasis was put on the demonstration of effect, with impact featured in both arms of the Dual Funding system, but mechanised most instrumentally through REF. Put simply, Canada focused on *how*, and the UK on *what*. In a further Poirot-esque realisation, we noted that in both systems there was a noticeable obscuring of the people who drive impact or use research (i.e. 'who'). Given that there is a no-one-size-fits-all solution for impact, we reasoned that doing impact well relied on the ability of people to comprehend and critically assess how their research could be best connected with the world. Not only that, but a deeper understanding of how impact works would also help people 'hold the line' when pressures start to weigh more heavily.

And just like that – *TA DA* – the concept of impact literacy was born.

WHAT IS IMPACT LITERACY?

Impact literacy is the ability to understand, appraise and make decisions about how to connect your research to the outside work. It reflects the understanding necessary to align the 'moveable parts' of impact to develop and execute meaningful, appropriate and realistic pathways to generate benefit in the 'real world'.

We'll come onto the elements of impact literacy in a moment, but the rationale for it is this: given the breadth of possible ways research can lead to impact, we need to be independently able to judge the most appropriate approaches in our own situation. In contrast to impact frameworks which focus on a sequenced logic of steps, impact literacy focuses instead on developing sufficient cognisance of the what, how, who and why of impact to configure appropriate pathways between research and effect. Impact literacy is not saying existing models are wrong; it is saying that 2D models are generally fine until you need to apply them in practice, at which point trying to follow a script written for an impact performance that isn't yours can swiftly bring anger and despair. If the models of impact are the music theory, impact practice is the art of composition, and for that we need to be able to understand the way different elements can fit together harmoniously. That's impact literacy.

EVOLUTION OF THE MODEL

In our first model (2017), we expressed impact literacy as the intersection of three elements (see Fig. 2):

This model represented impact literacy as a combination of *WHAT, WHO* and *HOW* (we'll talk about each of those in more

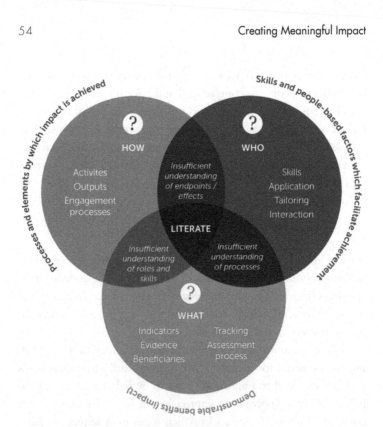

Fig. 2.   Original Impact Literacy Diagram.[1]

detail in a moment), but as we used the model in practice, we realised it needed three key extensions: recognition of the role of the
organisation (which we'll cover in Principle 5), recognition that
impact literacy can become more advanced (later in this section)
and a fourth aspect of literacy- WHY.

This led us to reinvent the impact literacy diagram (Fig. 3) to
combine the original components with these extensions:

The revised model thus includes the four components (WHY,
WHAT, HOW and WHO), reflects both individual literacy (right
side) and institutional literacy (left side), along with a sense of how
literacy grows (going up the pyramid). This chapter focuses specifically on individual literacy - the right side - and we'll cover institutional literacy later in this book.

[1] Bayley, J. E., & Phipps, D. (2019). Building the concept of research impact literacy.
Evidence & Policy: A Journal of Research, Debate and Practice, 15(4), 597-606. First
published online 2017, https://doi.org/10.1332/174426417X15034894876108

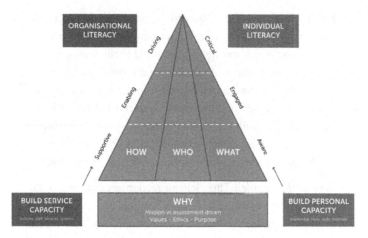

Fig. 3. Revised (Extended) Model of Impact Literacy.[2]

The final four core elements of impact literacy are:

WHY (The reason): The *WHY* of impact literacy reflects the rationale for embarking on impact at all. It's the answer to the question *why does it matter?* and is steeped in values and ethics about why it's needed, for who, and who says. Even though *WHY* was the last element to be added, it is actually the most important for setting us on the right mission. It wasn't intentionally absent in the first version of the model, we just had not quite realised how much the pace and wider challenges of impact often limited people's opportunity to fully reflect on 'what matters'. By bringing *WHY* not only to the fore but as the foundation of the revised model, we could signal how imperative it was to check this at the start.

WHAT (the changes): This element relates to impact goals and is the answer to the question *what changes?* As impact can't be templated, it becomes incumbent on us to be able to identify not only the impacts (changes), but also what problem(s) these address and appropriate indicators, measures and evidence we can use.

[2] Bayley, J. E., & Phipps D. (2019). Extending the concept of research impact literacy: levels of literacy, institutional role and ethical considerations [version 2; peer review: 2 approved]. *Emerald Open Research*, 1(14) (https://doi.org/10.35241/emeraldopenres.13140.2)

HOW (the paths): This relates to the practices and processes we can use to translate research into impact. It covers any methods by which research is connected to the non-academic world, spanning the continuum from dissemination through engagement to more active coproduction. It is most common for research to employ multiple methods to connect research to society, reflecting both the varied options we have at our disposal (e.g. business engagement, social media, training and consultancy) and the nature of the audience we're trying to reach.

WHO (the people): This element recognises that people are the active ingredient in making it happen, being knowledge producers, brokers, advocates, champions, dissenters, communicators, project managers, facilitators, leaders, administrators, trainers, policy-makers, users, customers, evidence providers ... people deliver the research and people decide if or how to use it. Impact needs the combined efforts and skills or people both within and outside of an institution, especially to ensure any plans are tailored to the context. The WHO aspect is analogous to an orchestra, with lots of different parts needing to work in unison to make beautiful music. Sorry if you were just a bit sick in your mouth.

RISKS OF TAKING A NON-LITERATE APPROACH

It is of course entirely possible to embark on impact without taking an impact literate approach, but this brings a number of risks:

- Considering HOW + WHO, without WHAT risks poor understanding of the impacts (changes), indicators, evidence or the relationship of the impact to the societal problem. *You miss the need and the goals.*

- Considering WHAT + HOW, without WHO risks poor understanding of the people, skills and tailoring needed to connect research with the context. *You miss the relationships.*

- Considering WHAT + WHO, without HOW risks poor understanding of the ways in which research can be viably and appropriately connected with society. *You miss the processes.*

- And any of these with an absence of WHY means that even if the plans are good, they might not reflect what matters. *You miss the point.*

In practice this can manifest in a number of ways:

- Poor clarity on the problem, and what impact matters.

- Pursuing inappropriate goals and/or missing opportunities for meaningful change.

- Not measuring the right things, or in the right way.

- Misjudging how ready society is for change.

- Annoying non-academic partners.

To be impact literate is to centre your thinking on real-world change and work backwards from there. By explicitly thinking about *WHAT, WHO, HOW* and *WHY*, we can align our attitudes and our actions to give research the best chance of making a difference.

> *'I get so tired of hearing people assume impact is something that will spontaneously happen if the research is good enough. You have to meet the people you are trying to reach halfway, translating your work into resources and ideas they can use and want to use. Think of it this way – pencils are good things to write with, they are cheap, easy to use and you can erase mistakes easily. But most people choose to write with pens. Why? Because they are widely available from work stationary cupboards (so effectively free at point of use) and are the cultural norm for adults who want to write things down. Pencils are for children and artists, so pencils are not the go-to choice for the busy adult-about-town. Good research can be like this. Even if on paper it looks to have good application in some contexts (and great potential for application in others), using it might need a significant change in habit and/or culture. Knowing about an innovation and its advantages won't necessarily change your behaviour. Providing information,*

standing back and waiting for the magic to happen isn't going to rock it either. You have to be ready to put the pencil (not just the article about the advantages of pencil use) in the person's hand right when they are thinking about writing something. That might need you to help them question why they never thought to use it in that context before, and you might need to be ready with counter arguments to the disadvantages of pencil use from die-hard pen users. That means stepping outside the academy and into the places where the pens get used. And you will need to accept that some people will likely always prefer the pen.'

Professor Clare Wood, Professor of Psychology,
Nottingham Trent University

LEVELS OF LITERACY

Much like research, impact is complex and context dependent, meaning there's always something new to learn. It's not as simple as being literate or not. In fact, we amended our original model because ignoring variations in literacy capabilities risked oversimplifying what it meant to understand impact, unintentionally suggesting that impact didn't need ongoing support or recognition of growing expertise.

To help explain how individual literacy can grow, and help you identity where you are now, Table 1 outlines literacy at three levels: basic awareness of impact and the associated wider body of insights, through a more intermediate level (where people are far more engaged in impact thinking), to more advanced critical thinking where people are proactively driving impact knowledge forward.

Whatever your starting level, it is the pursuit of literacy that matters; driving ourselves to understand impact better and engage more. That's what lifts us from reactive to proactive, and from process to meaning driven approaches.

SUMMARY

Impact literacy is the ability to understand the WHY, WHO, HOW and WHAT of impact. It is the ability to judge *why* change is

Table 1. Levels of Individual Impact Literacy.

Literacy Level	Description of Level
Aware (basic)	Aware of the evidence about impact practices and processes, understands there is a body of expertise, knowledge and tools which can underpin practice, but may not use or know how to draw them into practice. Likely understands impact at project (small scale) level
Engaged (intermediate)	Informed by and engaged with the evidence, understands there is a body of expertise, knowledge and tools which can underpin practice, knows how to draw on these and builds them into practice. Likely to be able to comprehend at a programme (higher order) level
Critical (Advanced)	Critically engaged with the evidence, understands there is a body of expertise, knowledge and tools which can underpin practice and is able to: (i) synthesise; (ii) critique; and (iii) add to/extend it. Likely to be able to comprehend at a strategic and/or systems level

Source: Adapted from Bayley and Phipps (2019).[3]

needed, *what* impact is being pursued, *who* needs to be involved and *how* this can be brought to fruition. Impact literacy isn't complicated, but more a call to arms to really focus on the different parts of the impact process and be able to make fair, appropriate decisions on what works best. Impact literacy is about using values and existing insights to inform our choices when there is a decisional fork in the road.

The WHY aspect reminds us to centre everything we do on meaning, and what matters to people outside of academia. Without really sense-checking this, we risk heading off in a wholly wrong direction. The WHAT aspect reminds us to be clear on the changes themselves, so we follow any paths right through to the point change(s) happen. Without this, we might be doing brilliant (or terrible)

[3] Bayley, J. E., & Phipps D. (2019). Extending the concept of research impact literacy: levels of literacy, institutional role and ethical considerations [version 2; peer review: 2 approved]. *Emerald Open Research*, 1(14) (https://doi.org/10.35241/emeraldopenres.13140.2)

things, but we can't reasonably know what's worked. The *HOW* aspect reminds us to really think about the methods which will *work* to connect with people, especially bringing them not only close to but *part of* the research to impact journey. Without this, we risk putting a hell of a lot of time and effort into activities which won't achieve what we need them to. The *WHO* aspect reminds us that impact is about connecting with people, without whose passions and skills we really can't do anything at all.

Impact literacy exists because whilst there is no-one-size-fits-all approach to impact, there are psychological, professional and financial penalties associated with 'getting impact wrong'. Rather than address this by offering prescriptive methods, impact literacy reminds us that the answers sit in collaborative, informed and goal focused practices to meet a real need. Impact literacy enables us not only to plan well, but to be responsive to changes in the messy world of real life. If our plans start to falter, we can assess which aspect of *WHY, WHO, HOW* and *WHAT* may need adapting. An individual is research impact literate if they understand the needs and benefits being sought, the activities which would achieve these, and who needs to be involved. Being armed with this inner sat nav helps you drive the impact car wherever it needs to go.

WHAT CAN YOU DO?

Whilst you're embracing impact literacy, embrace the chance to become *more literate*. There are several ways to build your literacy. Obviously one option is to get a book on it, which I would say is a most excellent idea. But there are far more options than that. It's important to recognise that whilst you might be new to impact, there is a substantive history of insights to immerse yourself in the world of knowledge mobilisation, research application and implementation, and impact. The wealth of research and scholarship on implementation is growing at a seriously impressive rate, paying testament to research translation is becoming ever more 'mainstream'. And insights for impact are everywhere; in the people who are immersed in the practice of impact, the research which seeks to understand implementation pathways, the researchers who try to drive their work forward, and the mission groups who are looking for change. Building your literacy is about opening yourself up to this world of understanding. It doesn't matter how new or senior you are, how fancy your research is or if you're required to create a formal case study. Impact literacy is about deepening your understanding of the way research can fit – like a puzzle piece – into what society needs, and do whatever you can to complete the picture. And given how busy we all are, impact literacy can help us focus our efforts in the most sensible and optimal way whilst we spin, drop and break all the other plates in our lives.

So to go from impact Padawan, to Jedi, to Jedi Master:

- Speak with 'real people' about what matters – businesses, charities, patients, artists, teachers, whoever in your world represents that part of life outside of academia.

- Speak with the impact experts and impact-active academics around you to draw on their tacit knowledge and experience, amassed by getting it both champagne-poppingly right and eye-gougingly wrong.

- Read up. Dive into the growing literature on impact, practice, implementation science, research-on-research and evaluation, and look for impact-related reports such as those on sustainable development goals, charity reports, funder's reports, case studies ... the list is endless!

- Get involved – join impact-related special interest groups, networks or communities of practice (academic, research management or wider community). There are commonly but not always cakes.

- Try things out. If they work, great, but if they don't just decide it's a 'teachable moment' and move on with the insights it brings. Impact is a learned art and what in which mistakes are made.

- Enjoy

> 'Don't panic. We've all screwed up something at some point.'
> Lorna Wilson, Co-Director of Research & Innovation
> Services, Durham University

- Do I pay enough attention to the *WHY, WHO, HOW* and *WHAT* of impact?

- How can I make sure I base what I do on an understanding of WHY?

- Have I thought through WHAT needs to change?

- Have I identified WHO I need to connect with?

- Have I considered HOW I might best connect my research with those outside the academic walls?

- How literate am I (on a scale of aware through to critical)?

- What opportunities can I use to build my literacy?

Chapter 3

IMPACT, VALUES AND POWER

Impact case studies show the sausages, not the sausage factory

Before we move on to principles in Part 2, I want to take a moment to dive into a couple conceptual and ethical issues which surround impact. I can't do full justice to these in this chapter or even this book, nor will I have covered everything that matters. My aim is to flag some key issues, and urge you, more than for anything else I talk about, to really engage in wider debates on these. We'll reflect on these throughout the remainder of the book, but it's important they have their own space too.

HOW WE PRESENT IMPACT SKEWS WHAT WE THINK COUNTS – *BIG, SHINY ENDPOINTS*

Have you ever read stories about the amazing benefits research has made – in impact case studies, university websites, funders pages, newspapers, social media and such like – and thought *wow*. Or made a mental note that 'this' is what impact is, and what you need to replicate in a funding bid. And/or felt a pang of inferiority?

Impact is, at its heart, making a difference through research. But certainly in assessment-driven contexts and those with formal agendas affording financial or reputational wins typically necessitate accounts of impact success at their grandest and most significant levels. Impact case studies are always going to show the wins, that's what they're designed to do. And there should be no scenario ever where we should mute, dilute or shy away from telling genuinely positive stories about the connection between research and society. Those stories are incredible and need celebrating. Seriously, go look at the case study databases, websites and impact assessment reports. The efforts, blood, sweat and tears to make those happen should never be underplayed, and anyone who can show real-world change from their work should feel the warm glow of an impact job well done.

The problem is that too often the effect of selecting and showcasing only-the-wins is that these high scale effects are cognitively (and systemically) bookmarked as 'what impact is', leaving people feeling inadequate when they can't mirror them. But if you think about it, case studies (certainly for assessment) are almost routinely:

- Positive (only showing the wins).

- Suggestive of a completed journey, positioning impact as a calculable 'endpoint'.

- Written by academics.

- Selected from a larger set of possible stories.

- Absent of, or at least minimising of, smaller effects.

Case studies are often used to represent, or provide a bedrock of analysis for, 'how' research leads to 'what types of' impact. But over-reliance on a data source already so filtered perpetuates expectations that impact runs along linear paths to what we're going to call Big Shiny Endpoints (BSEs) from hereon in. Setting our impact sat-nav to destination BSE immediately risks depriviledging studies or subjects which don't have a singular line of sight towards vast change, and masking any smaller (however meaningful) changes along the way. By screening out the full diversity of impacts and routes to change, we create schemas for 'what impact is' and become risk averse to chasing smaller effects which might not score so well.

More dangerously it can shift us to focus on what can be counted, rather than what counts. Doesn't seem the best thing to do, does it?

'*There is a common misconception as we seek to measure more of what matters about our research endeavours, that because we value impact, we should measure impact, and that measuring impact involves counting it in some way. Neither of these things are true. Trying to incentivise impact by measuring it can just leave us with a large volume of homogenous research as researchers focus only on the countable. And given that real-world impact is multi-dimensional and multi-directional, it cannot simply be reduced to a single, or even a series, of numbers. Sure, some dimensions of impact may be quantifiable, but not all, and certainly not all the things that really matter.*'

Dr Elizabeth Gadd, Research Policy Manager,
Loughborough University & Chair of the International
Network of Research Management Societies (INORMS)
Research Evaluation Group

WE DON'T TALK ABOUT FAILURE (OR HARM)

A few years ago, whilst walking through London with my husband, we passed Spearmint Rhino. For those of you who don't know, or don't want to admit you know, Spearmint Rhino is a fairly famous strip club, with branches (do strip clubs have branches?) in various countries. I – jokingly – asked him if I should apply for a job there.

Only if you like rejection, he replied, with a concerningly straight face.

Harsh, but fair. I of course have no real interest in working there – I'm far too unwilling to lose my University library access rights and Eduroam Wi-Fi privileges to leave academia – but it always leads me to think about the prospect of failure.

Impact is a far more complex, engaged and risk-filled process than case studies can bear witness to. However genuinely brilliant, such snapshots are like a glossy and unageing photo of a holiday romance, a sanitised version of a bigger and invariably messier affair. I'm yet to read a case study which says 'we tried something, it went wrong, so we all cried and drank gin'.

Failure is a central and incredibly important aspect of impact, yet routinely filtered out of impact discussions. Now when I say failure – which is perhaps too negative a term – I'm using it as an umbrella term to cover plans that are for whatever reason unrealised, altered, modified or unachievable, particularly:

• Plans which don't come to fruition because of circumstances outside of your control (such as the business that goes bust).

• The need to try different paths to see which 'work' (meaning some routes will be unsuccessful by design).

• Change in researcher circumstances which mean long-term plans aren't realistically deliverable (e.g. changing job, becoming a carer, receiving a 'with regret' email from a strip club …).

• Opposition and resistance within society, particularly in topics which are controversial or sensitive.

It is of course in no institution's interest to say 'we could've had this impact, but a load of stuff went wrong', but sector-wide camouflaging of barriers, misunderstandings, lost opportunities and ethical dilemmas not only gives us sector amnesia for how impact works, but also collectively implies that anything marred by difficulties isn't normal. I've seen too many colleagues convinced of their impact inadequacy, resolving themselves to the pointlessness of pursuing smaller effects, having internalised lack of impact as a failure on them, rather than a consequence of more contextual factors. Just as with publication biases against null findings, collectively muting 'what doesn't work' inhibits impact, dooms us to repeat misjudgements, and continues to allow individuals to mark themselves against an often unachievable benchmark.

The combinations of BSEs and screened-out failure also masks the real and dangerous possibility that research could cause unintentionally or negligently cause harm.

'It is all too overlooked within Impact culture that failure to achieve the right impact, is also impact. There are two parts of impact: the light and the dark. Currently many focus on impact literacy as impact that is linked to

academic prestige, and as a result we focus on the reward rather than how our behaviour, our research and partners operate towards unintended goals. We must remember that grimpact IS impact, and that it lurks, between our enthusiasm to seek reward for a job well done and our collective reluctance to admit that something went wrong and that we failed to benefit society. Our collective blindness to accepting grimpact as an integral part of the science-to-society relationship, is as dangerous as any form of blatant academic misconduct. But it is also not a mistake. It is an opportunity to reflect, to learn and to be wiser the next time an opportunity for impact comes our way.'

Dr Gemma Derrick, Associate Professor
(Research Policy & Culture), University of Bristol

'We cannot gloss over negative impact as we have been doing (and still do) with negative results in academic research. Negative impacts can represent real harm; we only need to look at the damage caused by the belief that vaccinations cause autism to demonstrate that point. Outside of such controversies (and yes, maybe the initial research was bad or the conclusions were badly communicated), sometimes it's unclear what the impact intention of a piece of work is, or if there was an intention at all. What is someone's intended positive impact (even with the best of intentions) might not be a positive impact for someone else, and what might be impact at a certain place in time might not be impact in another. We aim for impact to be positive, but we can't pretend research doesn't have the power to cause harm too.'

Esther De Smet, Senior Policy Advisor, Research
Department of Ghent University

WHO DETERMINES WHAT IMPACT IS?

'In Canada we have specific protocols and agreements that govern research with First Nations communities. OCAP stands for 'ownership, control, access and permission' which

means that it is the First Nations community that owns, controls, has access to and possess the research and its outputs. The community drives the research agenda and tells the research what is important. What counts is defined by the community not by the researcher. There are also clear overlaps between Equality, Diversity and Inclusion (EDI) and knowledge mobilisation (KMb). In Research Impact Canada (www.researchimpact.ca) we have started thinking about the intersections between KMb and EDI, driven by our observations that both EDI and KMb are essential elements of excellent research and concerned particularly with removing barriers and enhancing access to research. And this is demonstrated no more clearly than in our work with marginalised communities, where KMb and EDI cannot be disentangled. We are continuing to explore how we can support EDI as we support KMb in Canada's research enterprise.'

Dr David Phipps, Assistant VP Research Strategy & Impact, York University (Toronto), and Director of Research Impact Canada

Impact is, well, impact. It is the provable benefits of research in the real world, and on that basis the only people who can determine what counts as impact are those who are affected by it (or represent the aspects of life that are). But that is of course an almost naïve stance given that impacts are also often formally judged within academia, be that within funding bids or in assessment.

When it comes to who *formally* determines what impact is, then it's easiest to think of this as agenda-setters taking the core point of impact and add a layer of *what counts for them*. Funders commonly seek impact within their remit, such as environmental impact for environmental funders or health impacts for health funders. For assessment, impact typically 'counts' when it satisfies a range of eligibility criteria around timescales, staff employment, research quality and disciplinary 'fit'. There's obviously flex all around this, but these equations (funder = impact + remit; assessment = impact + eligibility) give us an insight on the lens through which impact expectations are variously shaped. And all that's before there is a judgement about what of this is (competitively) *better*.

I can't leave this part without a cautionary note about not letting the formal agenda tail wag the impact dog. Nor can I move on without commenting on how impact has become a loaded term, both by conveying a sense of 'collision', and in its worst form positioning research as a kind of one-way super gift in which cleverness inside the academy translates into a grateful world, sat waiting to be lifted from its problems. There are, of course, examples where this is essentially true – we only need to look at the example of the COVID vaccine to see that – but in wider practice the truth is far more complex. Society is rarely in hibernation, and is instead gloriously active with people trying to make a difference in a million different ways. If we position research as *the* saviour, we risk infantilising and disempowering wider society as a kind of empty vessel for our ingenuity. Thankfully most people are really good at not thinking like this, but if we put together the pressures of academia with the systemic picture painted of what impact 'is', we can see why people might fall into that trap.

So who determines what impact is? It depends who's asking.

'I was in a situation writing an impact case study for a researcher. He explained that he'd been present at an event where other researchers made a decision that had sector-wide implications, and claimed this as his impact. It turned out he'd not actually contributed to that discussion (which seemed actually to be just a coffee break), and was simply 'near' those who did. I advised him that this couldn't be claimed as impact, but not only was he was adamant that this was his impact, he argued that he deserved to be rewarded for it. We know that impact is much broader than we often recognize and in many situations we are 'impacting' without realizing it. But as in this case it is all too common to misinterpret what impact is, and to justify doing or claiming something, 'because impact'. The truth is that impact can't just be claimed without sufficient grounds, nor is it impact until someone says it is. Unfortunately the system perpetuates pressure to find impact anywhere, and at any cost, meaning people

rush to see impact on everything that is beyond academic effects. But really, those who define it as impact are those who evaluate it.'

Dr Gemma Derrick, Associate Professor
(Research Policy & Culture), University of Bristol

PRESSURES ON PEOPLE AND INSTITUTIONS: LABOUR, VISIBILITY AND SURVIVALISM

'Impact assessment, done irresponsibly, is like a strip club. Some people come in with money, some leave with money, and everyone feels a bit dirtier.'

Me, childishly, in various conference talks

Institutions can get a bad rap when it comes to impact, particularly when it looks like all they're bothered about is assessment, case studies and BSEs. But for many institutions in some countries, there is no option *not* to do this, or to focus scarce resources on the thing that's going to help them be most financially sustainable. We'll come to institutional health/ill health in Principle 5, but it's important to remember that impact is an increasingly weighted requirement for funding, both through competitive bidding and in assessment where it exists. As a result, failure to deliver 'good impact' can have both financial and reputational penalties, with institutions who do not perform receiving little or no allocation of money. Survivalism is no small driver for impact, sad as that is.

It would be extremely easy to launch into a monologue at this point about disproportionate resourcing across the research sector, and how many people feel very burned out with little reward. But I'll simply say this: extending my strip club analogy further, it's not uncommon to feel – accurately or not – that others are getting £50 notes stuffed in their impact pants[1] whilst the rest of us are on bar duty.

Stepping away from this pants-based-money envy, there is also an ethical dimension intrinsic to everything we do. We are often

---

[1] And yes, I mean the UK version of pants here.

publicly funded, and society quite reasonably expects us to return their investment in a way that's helpful. That doesn't mean all research should have impact, but rather that it's not at all unreasonable that we should seek wherever possible to deliver on our civic commitment to society.

> 'Mike Daly (Dept. Earth and Space Sciences, York University) studies exo-planets and the origins of the solar system. He is part of a multinational project that sent a space craft to an asteroid named Bennu. It is returning to earth, due to land in 2023, with some material from Bennu. It is an asteroid return mission. How COOL IS THAT? Sure, his research will never inform a new public policy or provide a needed social service. But he makes presentations to school aged kids in museums. His outreach not only imparts new knowledge to kids but might turn on new minds to space engineering as a career. Now that is also impact.'
>
> Dr David Phipps, Assistant VP Research Strategy & Impact, York University (Toronto), and Director of Research Impact Canada

Inherent to all this however is a tension around what impact means for 'being an academic'. For so long, and still in many areas of the sector, more traditional models of academia have prioritised income, outputs and indicators of esteem. Suddenly this impact agenda whistles its way in and demands that we also now get cosy with society and 'make a difference'. For many that can feel not only unnerving but threatening. I find it's generally rare for people to be anti-impact, instead being anti-formal impact *agendas*, commonly reflecting three annoyances:

- Why aren't the contributions I already make to academia enough? Or *I do a LOT of research/teaching, why do I need to do more?!*

- Why aren't the contributions I make outside of research (e.g. teaching) counted? Or *gees stop looking at things so narrowly.*

- Why do I have to prove it? Or *don't you trust me?*

Experience suggests some people are also unnerved by the prospect that wider collaboration could undermine research quality, or introduce the need to compromise their values in some way.

'For academics, particularly in arts and humanities, knowing how to relate to impact in an authentic way can be incredibly challenging. Sometimes there's no clear route to impact, and sometimes we might find ourselves asked to deliver research or communications which step beyond what feels right or crosses a line. It can be uncomfortable to push back in those circumstances, especially when there is a risk of compromising partnerships and collaborations, but actually people respect academic integrity and respect us more when we are clear on what we feel is acceptable. I urge everyone to figure out their personal 'red lines', and don't worry if you cross them as sometimes that's the way we find out where they are.'
Professor Ele Belfiore, Professor in Cultural Policy & Director of the Interdisciplinary Centre for Social Inclusion and Cultural Diversity, University of Aberdeen

The sector is still working out how best to balance 'traditional' aspects of academia with impact, not only within the university, but across the research ecosystem. Funders are increasingly recognising the need for impact-related resources, and more routinely providing guidance and awards for public involvement and impact. Publishers are increasingly stepping forward to help innovate and amplify research messages to reach key audiences. Institutions are developing new ways to better integrate impact into processes, such as creating secondments and sabbaticals. And impact is increasingly being valued in progression across the system. But we need to keep going to really embed it in – not just add it too – what we do. If not, all the motivation in the world won't stop people getting worn out.

'Dementia care is about understanding identity and what matters to people. It is about understanding how people and environments can empower or disempower, enable or disable people living with a cognitive impairment. My work centres on the psychological wellbeing of people via

a method known as Material Citizenship. This focuses on the importance of everyday mundane objects, such as a certain coffee cup or curling tongs, and how they can support a person to live the life they want to live, their way. Objects also provide a mechanism for maintaining and demonstrating identity and in turn the ability to develop deeper relationships between people. What's been so frustrating for me personally, in my academic career, is seeing the huge potential for research to actually make a difference cut off by systems and processes. In essence the research sector gets in its own way! It can be so immersed in what an academic needs to look like, and what papers and grants they need to produce, that they can lose sight of the benefit of the knowledge generated and the difference this can make to people's lives. Career progression prioritises metrics of income and publication, but whilst people might feather their caps with fancy journal articles, whose lives are changed? What is the point of us gaining knowledge if we don't put it into action? So often people living with a dementia and those who care for them engage in research often not for themselves but for others who might benefit further down the line. It is fundamentally unethical knowing this, not to use the knowledge we generate to make a difference to those it is intended to serve. Unlike the healthcare system, social care lacks an R&D infrastructure, change might be less tangible, and staff don't have the same standardised opportunities for training. This can make it far harder to make a difference, but not impossible, as I have shown. It is for academic institutions and funding bodies to adapt and understand, to support, to reshape it's approaches and to recentre itself on the world outside its own script. For me, the only real option was to step out of academia to make the difference I knew was needed. The blockages caused by pressures to write (rather than do), protracted intellectual property negotiations, and the ongoing gulf between what we know and what we change became too lengthy. It was important for me to see my research change the dementia care landscape, after all that is what I set out to do. I also have a need to see this change in my lifetime. Whilst others might not feel such a strong pull, it's extraordinarily clear to me that we need a full reassessment of the fullest scope of academic contribution, the ways we demonstrate it, and the way in which we genuinely value, not just pay lip service to, changing the lives of real people.'

Dr Kellyn Lee, CEO and Founder WISER Health and Social Care

The downside of an agenda amplifying the need to increase the wellbeing of society is the risk of being ignorant to (or worse apathetic towards) the toll on those *within* academia. Because weirdly enough, researchers are people too. Impact is rarely the effortless result of excellent research 'existing'. The efforts to drive it towards social change sit alongside the other pressures we already face in academia – publish or perish, job precarity, chasing funding, teaching schedules and many more. With disproportionate challenges for colleagues who simply don't have the bandwidth to do 'extra' (let's be clear there should be no expectation to do extra anyway, but it is particularly marked for those with additional pressures such as ill health or caring responsibilities), increased workload and extended maps of professional success risk deepening and cementing pre-existing inequalities even more.

Impact requires the efforts of many people, but accounts can neutralise the contributions of those outside the core research team, such as other researchers, methodologists, students, professional services, librarians and so many more. This risks not only suggesting the route is effortless, but also polarising teams between highly visible 'star players' and hidden 'support staff' who struggle to have their efforts recognised. If we don't 'unhide' the full team, not only will we never convey the full picture of how much combined effort is needed, but we might as well just print 'YOU DON'T MATTER' stickers and hand them out en masse.

'Look around your workplace and recognise that your work is more diverse than appreciated in research assessment frameworks. We created the Hidden REF programme to celebrate the efforts of so many people, and so many contributions from research, that are made invisible by our assessment

processes. Our publications and our impact are not the only outputs produced by science and our lab tech, librarians, research managers and research participants all contribute to our research, and yet their contribution is overlooked. We were in awe of the work displayed in the nominations to our 2021 round, and the winners within each category demonstrated the sheer scale of commitment and expertise across the sector. It's sad they're so often hidden from view. The more we can do to recognise their efforts, the more accurate the value of research culture is perceived.'

Dr Gemma Derrick, Associate Professor
(Research Policy & Culture), University of Bristol

MECHANISING RELATIONSHIPS

We know that in impact everything is improved by coproduction, from understanding the need, through research design and into implementation and impact itself. We also know that involvement and engagement of any kind gets us far closer to what matters, how our research can contribute, and the ways we might be able to demonstrate that change. But that doesn't mean we always do it well, even when we mean to do it right. Just as we risk hiding efforts within academia itself, we also risk making our external relationships instrumental, established only to tick a box or provide us with evidence. And that has a tendency to really annoy the hell out of the people who give up their time.

The world of Public and Patient Involvement (PPI) gives us some pivotal insights here. PPI is fundamentally important to healthcare research, and focuses on doing research *with* people, rather than *on* them. PPI is increasingly required in a lot of healthcare research, and patient advocacy more widely has become an increasingly expected and accepted part of both research and care itself. When done well, PPI is wonderful. Patients have equal voices in the process, co-designing, co-delivering, co-interpreting and co-reporting research which reflects patient-identified needs. But PPI can also be done very tokenistically, where patients are superficially included or used purely to 'sign off' bits of the research already designed. The effect for patients is not to feel involved, but to feel *used*.

Some areas of research of course lend themselves far more to wider engagement than others, but that doesn't mean we should take any less care about the relationships we do build. Trust is an extraordinarily important aspect of impact, as without it how can we ever expect people to want to use the work we produce?

'I was involved in a significant piece of work a few years ago, designed to explore and get political traction on the importance of cultural value. We completed the work, published a report which was then publicised to great fanfare across the media, and mentioned in the manifestos of various parties. But then the fanfaring – and plans for a case study – only focused on the fact the report existed and had been cited in the press and had shaped debate. But, in my eyes, the real impact was how compellingly the report exposed inequalities in cultural participation across the country. We had wanted it to be a report supporting social justice, but the points of embarrassment became hidden from view. It felt that the inequalities were being ignored as the university looked for the 'shiny' case it could narrate, compounded by the fact that efforts to keep engaging with stakeholders 'stopped' once it was published. We felt we had been shortchanged and we felt even more that the trust the cultural sector had given us had been broken. Thankfully, and because we had invested so much genuine energy in building partnerships, colleagues in the cultural sector working at the coalface of accessibility and diversity made us aware of the benefit the report had anyway, in feeling seen, being seen as legitimate and allowing them to articulate their value. So whilst this experience shone a light on how we can easily end up paying lip service to things, even when we never intended to, true and trusted partnerships will not only endure but also remind us the impact wasn't necessarily where we were looking.'

Professor Ele Belfiore, Professor in Cultural Policy
& Director of the Interdisciplinary Centre for Social
Inclusion and Cultural Diversity, University of Aberdeen

RECOGNISING PRIVILEGE

All this said, we also need to recognise that there are privileges within academia which are not always afforded to others in society. Notwithstanding workload issues, we do have the opportunity to deploy a range of theories, models, skills, paradigms and all else to help make a difference in a way that many outside of academia can't. And we also have a voice. It may not always feel like it, but we have the chance to shape learning, work with experts, connect with peers, apply for (and if lucky successfully obtain) funds, design learning materials, set curricula, create student opportunities, write papers, write blogs, write books, be part of project teams, write strategies ... that's not suggestive that such outlets are *only* the domain of academia, but it would be wholly disingenuous to suggest these aren't some of the perks. Certainly for many of us, the question is how we can use our actions to elevate the voices of those who aren't as easily heard.

'Just the joy of knowing and working with Julie[2] and contributing to this book, is a product of compounded privilege. My contribution is here by the merit of having a full-time, permanent academic position, as well as having the funds and support available to participate in conferences, and projects where I met Julie (and other inspiring researchers) and being able to capitalize on these connections over time. So, while I am enthusiastic about the beauty of working in research, I acknowledge that I am here not just through hard work, but also by benefiting from a system where access is governed by processes that advantages some, and disadvantages others. I cannot forget this as I work to improve research culture, and I would like to think that all researchers would do the same. After all research is enhanced by many voices – some we don't like to hear, but we can still build better research, and societal outcomes, by listening to them.'

Dr Gemma Derrick, Associate Professor
(Research Policy & Culture), University of Bristol

---

[2] *I promise she said this without being paid.*

TOWARDS FAIRER

I've tried in this chapter to outline some of the issues of power within impact, but there's a risk this feels wholly dispiriting. And it won't surprise you to know it reflects many of my own experiences and disenchantment at times. I'll always champion impact, because helping the world matters, but I can't do that without recognising pressures within the system that can corrode the motivation of even the most impact-jazz-hand amongst us. I gravitated to impact because I was so disheartened that the interventions I was designing didn't really seem to make a dent in the world, and have since despaired far too many times trying to swim in the BSE-flowing river. The two things that draw me back, every single time, are the passions of colleagues so joyously committed to their work, and the constant possibility that we can make a difference in a way that matters. Once you see the devotion of researchers to their topic, and the commitment of people outside of academia to their realm of life, you cannot fail to be reminded of how important that little old word 'impact' is. And I am reminded of the privilege of being able to help support people to do impact in every conversation I have.

But you don't need my life story so let's move on.

Thankfully there's already considerable positive movement in many of these issues outlined in this section, and so these issues are perhaps far more reflective of an agenda still trying to fully find its feet in a sector already awash with a bunch of challenges. How do we do it? As it's not in the interest of institutions to shout about what goes wrong, nor by extension would academics necessarily feel safe to 'admit to their failures', we need to collectively agree to take a wider view on impact. It's unrealistic to think that anytime soon page-limited case studies will be intoned with the inherent messiness of impact, nor will limited funds be made freely available for things that won't work. So instead, practically, we need to fly at this from a number of directions to build literacy, reduce pressures, acknowledge the messiness of impact and strengthen our overall connection with society:

• Value impact of any size, rather than dismissing smaller effects or those which have a more scenic route.

- Share failure: Make it not only safer to share 'what doesn't work', but make doing so a legitimate and valued part of academic practice.

- Recognise and appreciate the time needed to build trusted relationships, to avoid oversimplified assumptions that research will naturally 'translate'.

- Recognise and address the pressures of adding impact into an already busy environment.

- Collaborate, coproduce and recognise effort: true dialogue and engagement of the sort that is as far from tokenism as it's possible to get.

How do we do it? We need to work together.

'Impact has at its heart a very simple principle – to make a difference through research – but as a sector, we have conformed to a much narrower definition of impact for far too long. Our focus on impact within the academic community, rather than outside of it, has narrowed our view of what constitutes 'quality' research, driven by appetites for citations and the Impact Factor. We can't shy away from the fact that some people and organisations have benefitted from this, including publishers. It has created a hierarchy, a benchmark, a standard, a way of assessing, but that takes us away from how we can make a difference through research. Can we do that if we have such a narrow view, if we perpetuate harmful views of quality and assessment, if we only hear the same voices? By its very nature, research should be inclusive, but our sector rewards according to rankings and output which favour certain demographics and promote an elite system. I'm firmly of the view this means we're missing out. Amplifying voices from all parts of the world, from indigenous communities, from all protected characteristics, creates a much richer experience for us all. This will not affect quality, but it will upturn all of our deeply-ingrained perceptions of what quality is. We need different modes of conducting and communicating research. For publishers, this means thinking critically about how we

*help to drive change – publishing different research assets,
and finding new ways to communicate research beyond the
traditional article. It's also about ensuring that we reflect a
range of voices, which reach the broadest and most diverse
audiences. It's buoying to see some individuals and organisations trying to drive change, but they're still in the minority.
This effort needs to be systemic and sector-wide – we have
global challenges which require a sector-wide response, but
we also have a collective responsibility to promote healthy
research practices which support a range of research careers
and publishing opportunities.'*

Vicky Williams, CEO, Emerald Group

SUMMARY

This chapter probably seems really rather miserable, but that's
because we've needed to wrestle with some significant ethical
and power-related issues. It is essential that we – collectively as a
sector – invest time and thinking in these issues around values and
power, so that we can rebalance what it is to *do* impact, not just
achieve the biggest shiniest impact we can. But we need to find a
way to do that without pausing our efforts because society can't
wait whilst we argue amongst ourselves.

Impact is far more a work in progress. It's always fluid. And
the case studies are often the nicest story we can tell from what's
a much bigger, messier thing. Understanding failure, and barriers, and – as in my strip career that never was – the likelihood of
rejection, is as important to impact as understanding the benefits
research can bring. So I'm sticking with the term failure as a simple
catch-all, and in the hope that this helps us dilute the power of the
word in our academic lives. Let's own failure.

I don't want this to be a take-down of formal impact agendas in
the sector, as both funding and assessment operate as double-edged
swords; galvanising the sector towards societal benefit whilst creating a
new strand of systemic expectation. I feel duty bound to raise the challenges formalised agendas bring, but unbinding us from them wouldn't
necessarily serve us all any better either. Perhaps 'the jury is out' on that

one. For many of us though, the appetite for impact far outweighs the impetus of formal agendas, and what we seek instead is balance, fairness, insight, realism and meaningfulness. How we choose to use the privilege of academia, and how we – as a sector – galvanise action to address the challenges, is down to what we think matters.

My stance, for what it's worth is this. Research has the potential to benefit society and *where it can it should*. To be frank, it would be somewhere between insane and arrogant to think everything we do within research will land outside of the academic walls (who do we think we are? Steven Spielberg?). We also need to protect space for research which is more exploratory, and value learning from research which is 'unsuccessful'. And whilst we're doing that we need to take a damn good sector look at ourselves for the narrow way we operationalise what a 'proper' academic looks like.

> '*What we have to remember about the discovery of penicillin is that it happened because Fleming was a naughty boy and left the cultures out on the table when he went away for a few days.*'
>
> Derek Stewart OBE, Patient Advocate and Honorary Professor at the University of Galway

I am personally very optimistic about where we're going with impact, having been bowled over by how far we've come. These are of course my views, rather than an objective fact, so you're welcome to disagree. But as this is my book, I get to share my thoughts, which gives me an advantage in this particular regard.

WHAT CAN YOU DO?

Be reflective but don't despair. Take a step back and consider how you've built your sense of what impact 'is'. Is it imbued with expectations of size and scale, or reflective of the attitudes of those around you? Are you sold on impact or are you resistant about how it affects your professional identity? Are you visible? Are you hidden? Are you hiding others? Simplify your thinking: impact is the provable benefits of research in the real world, how can we set aside the distractions and do it in the fairest way?

- If I had to describe what impact is, would I include expectations of size and scale?

- If so, how have I come to understand that?

- Are BSEs helping me look for the right goals, or are they masking other ways I could make a difference?

- Are some people in my impact story in the shadows?

- How can I build reasonableness, fairness and inclusion into what I do?

Part 2

EIGHT PRINCIPLES FOR DEVELOPING AN IMPACT-LITERATE MINDSET

Principle 1

CHASE MEANING NOT UNICORNS

Impact is in the eye of the beholder

'*I was helping a small research charity to develop a new research strategy. Part of this work included stakeholder interviews to help define what success from research funding looked like for their perspective. Amongst the responses, the one which stood out for me and I will always remember was from a family with someone that had the rare condition the charity supported. Their 'lens' was that the health system had little understanding of their lived experience. For them, impact from research was not necessarily a cure but, as importantly, getting respect and understanding in their care, particularly being treated for their condition rather than confused with another illness. Always remember, their impact, their 'lens', their lived experience.*'

Dr Mark Taylor, multiple sclerosis patient advocate

What's an Impact Unicorn? Let me explain …

Impact is the provable benefit of research in the real world, but benefit is a subjective term. What is beneficial to me may not be to you, and whilst there's some obvious 'wins' (e.g. curing health problems), other effects might be definable only from a specific standpoint. For instance, increasing the market standing of one company may ultimately cause a loss in profits for another, and policies made in favour of one thing (e.g. sex education) may be wholly opposed by others seeking policies which block it. I once had a stand up row with a vicar in an ethics meeting because apparently research on teenagers' use of contraception 'shouldn't be needed'. Because we all know 'shouldn't' is the primary criteria by which teenagers make healthy decisions about sex.

As outlined in far more detail in Chapter 3, a consequence of the judged and essentially competitive nature of both funding and impact assessment is that our beliefs about 'what counts' get skewed. Assessment led approaches can fuel a tendency for people to prioritise BSEs, with paperwork favouring grander and neater stories, and funding applications fuel the sector appetite for promissory notes to fix big problems. Let's be clear – driving significant levels of change via policy, strategy or anything other 'big' way is phenomenal, and where we can find ways to benefit widely we should. The problem is with locking our impact sat-nav to 'destination big' can blinker us not only to smaller changes which can accrue (and be measured) along the way, but also convert our impact thinking into 'what counts' not 'what matters'.

In much the same way as I believe I have the singing voice of an angel, these BSE standpoints can too easily fuel *fantastical* thinking. Impact becomes a mystical, mythical and magical beast – an *Impact Unicorn* – where instead of focusing on the actual changes we can make (or those we can get on a path towards), we lock our targets on prestige. Unicorns exist when we put glory before what matters.

Once you lock on to the notion of Unicorns you can see them everywhere, particularly in actions such as:

- Discussing and supporting impact *only* in relation to the production of assessable or fundable materials.

- Focusing all impact efforts, processes and communications on trying to make a small number of case studies as high scoring as possible.

- Dismissing, trivialising or being advised against pursuing smaller impacts because they won't score as well.

- Prioritising impacts which make us feel less nervous about assessment rather than favouring those which count most to stakeholder.

- Over-promising in competitive bids.

Practically some of these things need to be in place, especially if you need to deliver impact for assessment, it's when they are the *only* or *most dominant* processes you're into Unicorn territory. And once Unicorns are there, they risk becoming deified and lauded, triggering cultish sector pilgrimages to possess them as a precious shiny thing. Ok that went a bit Lord of the Rings but the point stands. BSEs can be great, but when process turns into Unicorn hunting we've really rather lost the 'impact-is-about-what-matters' plot.

For a real life example which has stuck with me for years ... anyone who has been privileged enough to meet the incomparable and wonderfully kind Derek Stewart OBE (Patient Advocate and Honorary Professor at the University of Galway), can only ever have been struck by how meaningfully he talks about patient engagement and involvement. Derek is a throat cancer survivor and gives spellbinding talks about the need to centre the patient voice in research. I remember vividly sitting down for the first talk of his I ever heard, and suddenly realising 40 minutes later I hadn't even moved my pen to my notepad, such was my awe. And in that talk he said something that has stuck with me for many years, that when he was in hospital, and the medics were talking over him about treating the cancer, what he desperately wanted (obviously as well as the cure), was to be able to *swallow*. The simplest and most automatic of actions the rest of us would so easily overlook. Yet he hadn't been asked what he needed. How many times do we similarly presume what's needed without asking?

If we apply this principle to impact the danger is fairly obvious. Assuming, rather than checking what the problem is and therefore what 'benefit' would be can start us on a wholly wrong path, driving our limited energies into things which may not really make a difference whilst chasing big Unicorn'y versions of them anyway. Sure we might be able to measure them, but what have we really achieved? Academia is not, and never has been, the landowner of impact. Societal change will always be a collective act.

'Public and patient involvement (PPI) is fundamental to doing good research. It's about doing research with, not to people. But to some extent we've divorced the values of PPI from the practice of improving healthcare, and turned it into a transaction. PPI is a process, not an answer. It's about embedding what matters not just being seen to engage or tick a box. We need to redesign our landscape. Instead of a fenced off area we need to think in terms of commons land, an open space where everyone is heard and there is no sense of trespassing. In this way we build trust, and rather than just chasing big wins, bring research to bear on decision making and resolving clinical dilemmas. The people on that field don't all need to understand every part of the research process, but enough to bring their consid-ered view. The closer we get people with different skills and perspectives together, the more effectively we work out what matters and how we can collectively solve it.'

Derek Stewart OBE, Patient Advocate and Honorary
Professor at the University of Galway

ARE UNICORNS A PROBLEM?

My son recently told me that his class had been challenged to read the highest number of pages they could over the course of a week, with a reward for exceeding the previous week's total (I should say they have all sorts of reading challenges, not just numbers). I asked him how it was going. 'Great' he said, 'We all just decided to read graphic novels instead of full story books as we can whizz through the same amount of pages without all those words to deal with'. Can't fault 10 year olds' logic, but if that isn't a case in point about how people reshape their behaviours to fit the shape of the competition, I don't know what is.

Now you could of course say 'so what?' to the Unicorn problem. That we absolutely should be chasing national/international policy and that *only* change at that level is important. Firstly if you think that, then it's probably not worth you reading too much further as you're gonna hate the rest of the book. But my more professional response would be this. To presume only big changes – such as those at policy levels – count, wholly overlooks the myriad of ways people's lives can be made better. How can effects at any level *not* count? Tell that to the care user who wants to feel less stigmatised. Or the local museum who wants to avoid shutting their doors. Smaller wins matter, both because they can lead to bigger ones, and because *they just matter.* And call me cynical, but changing a policy doesn't always translate into better lives for normal people ....

I am always minded here of an experience when consulting at a university-which-won't-be-named. A senior academic was explaining his work as we discussed its potential for a REF case study. His work was fantastic, with clear routes to impact, but it became increasingly clear there was something wrong. He looked defeated, and when I queried why he was so negative about his genuinely brilliant work he started to sob, turning his computer screen to show an email from his boss:

> *It's all well and good changing national policy, but unless you change European policy it doesn't count.*

Well excuse me, but the last time I looked it most certainly did. To see someone so clearly capable flattened by such a Unicorn lens on impact, expressed by senior colleagues as an unquestionable *fact,* was heart breaking. The potential for a further type of impact doesn't simply cancel the one you've got. They're both impacts. Whether they count is fully down to the narrowness by which they're benchmarked. **Eye roll**

HARNESSING UNICORN ENERGY

If Impact Unicorns are fantastical, mythical and far off hunted beasts, our best response is to set ambitions which are realistic, achievable and meaningful for society. And it's ALWAYS best to work with stakeholders from the earliest point possible to avoid conflating impact paperwork with change that matters.

Unicorns aren't intrinsically bad, as they may often represent ambition. And if we're careful, we can harness Unicorn energy to fuel our aspirations, our efforts and our scale of contribution. As long as we avoid blinkering ourselves to BSE's alone, recognise that society generally doesn't want us to promise unachievable fantasies but rather crack on and do something useful, and acknowledge that narrow, deep effects count just as much as big fat numbery ones, we can grab that Unicorn and fly.

SUMMARY

'Sometimes the pressures from funding and assessment can push us towards tokenistic approaches to impact. We need to all try and push back on that and remember what matters to society. Starting with a sense of the impact and working backwards can often help us design stronger research and contribute more powerfully to the world around us.'
Lorna Wilson, Co-Director of Research & Innovation
Services, Durham University

Formal agendas can have a habit of making us look for magically impressive impact (AKA Impact Unicorns) which *can* divert us from what's meaningful. Unicorn thinking can make us judge anything that isn't big and fantastical as somehow weaker, less impressive or just *not normal*, which is of course nonsense. If scale and meaningfulness align, fabulous, but if they don't what should you do? The answer from outside of academia is simple – unless it's meaningful it's meaningless. Putting measurable ahead of meaningful turns impact into a tokenistic glory grab which serves only the academic institution. Case studies become competition entries rather than accounts of actual benefit, and relationships become instrumental rather than purposeful. And people within and outside of the university feel understandably screwed over.

I'm not saying we should only look for small effects – that's not the point at all. It's about not *only* looking for big shiny wins which blinker us to the wealth of opportunities to make a difference. Unicorns are a problem when they divert our choices, much like a child choosing a faster-to-read graphic novel over something more packed with words. It's vital that we address how the wider system reinforces Unicorn chasing, but we also need to be practical. For those needing to demonstrate impact, and for whom institutional financial or reputational sustainability may be at stake, it may be too risky *not* to have impressive things to measure. So absolutely work out what can be tracked, counted, testified to and generally proven, but don't let it *lead* your efforts. Moreover, don't confuse Unicorns with what impact is.

Thankfully meaningfulness is at the heart of many people's efforts, and whilst Unicorns hover in our peripheral vision, we are getting increasingly good at averting our gaze and focusing on the actual world around us. If you're still unsure how to avoid Unicorns, start with what matters, not just what's impressive, and you're already on the right path. After that, feel free to chase all the Unicorns you want.

And remember when you get up close, that unicorn is probably just an ugly horse in a pointy hat anyway.

WHAT CAN YOU DO?

If you think you're chasing Unicorns, don't beat yourself up. Reset your thinking using this elaborate 10 points strategy for establishing meaning:

1. Ask people

2. Ask people

3. Ask people

4. Ask people

5. Ask people

6. Ask people

7. Ask people

8. Ask people

9. Ask people

10. Ask people

- Have I put Unicorns above meaning? For example:

  o   Prioritised what can be documented over what should be changed?

  o   Prioritised what sounds impressive over what makes a difference?

  o   Decided against an impact opportunity because it isn't 'worth' enough.

- If no, fabulous, keep going.

- If yes, reread this chapter, take the Unicorn out of your crosshairs and look around. What matters is probably right in front of you.

Principle 2

WORK OUT WHAT YOUR RESEARCH POWERS UP

Given energy to the baton pass

During a hospital inpatient stay I had to do a pregnancy test before surgery. Negative results meant surgery went ahead, after which I became fairly unwell and was moved to high dependency, strapped down with wires and tubes. I shared the bay with a lovely octogenarian called Colin, and we played 'oxygen level bingo' to pass the time. Some more complications arose and I had to be whizzed down for 'redo' surgery, again being asked if I was pregnant. I advised that I'd tested negative two days earlier, and that I'd been tethered to a high dependency bed under continual monitoring since. I have to admire the protocol compliance of the doctor who, unable to just accept those results, asked 'yes, but has anything changed?'. I said that unless Colin's sperm had super powers and could jump 20 feet across the room, we were probably fine[1].

Lesson learned? Applying something bluntly without attention to context is a daft idea.

---

[1] And yes, this was the surgical incident resulting in the accidental drug-induced fake underwear purchasing saga in 'About the Author'.

In the pursuit of BSEs[2] it can be easy to overlook or disregard the impacts along the uncertain flightpath of research entering the real world. Yet it is in this 'along the way' that the true picture of impact emerges. The journey to change is often a chain of events and contact points which progressively connect research and society. So whilst it's absolutely fine and indeed advisable to set high level impact goals for your research, visualising the journey as a series of links can help you see how your research unlocks possibilities for those outside of academia.

In short, you can identify what your research *powers up*.

Let me give you a personal, non-research analogy. I'm a dementia carer, and like many carers, my input 'powers up' my dad's safety, dignity, access to social and health care and the likelihood of him being able to stay at home for the remainder of his life. My input *doesn't* cure his condition, improve it, or eliminate the growing scale of risks, but it makes his journey safer and less isolated than it would be otherwise. The 'impact' is the difference between his life with, versus his life without, that care.

Translate this analogy to your research. What would be different in the world if your research didn't exist? Mentally subtract your research from the picture and see what would be impossible, weaker, remain locked or otherwise be a problem. The difference is what you power up.

Powering up is the act of mobilising the relevant parts of your research (*WHAT*), in appropriate ways (*HOW*) to those who can use it (*WHO*) for a reason that matters to them (*WHY*). Let's break this down.

WHAT CAN BE MOBILISED?

'Research' is actually a bit of an umbrella term when it comes to impact, masking the suite of different findings, outputs, methods, insights and 'things' that can each fuel a route to impact. If we put our *powering up* hat on, it allows us to look less at research as a block, and more as a cupboard full of goodies, each of which

---

[2] Big Shiny Endpoints

may suit a different person or situation. Research might produce, for example:

- New knowledge, insights or evidence (*'we now know'*).

- New concepts, ideas or perspectives (*'we have rethought'*).

- Insights into ways things are understood (*'we know how we know'*).

- New or modified research methods (*'we know how to'*).

- Experiences of what doesn't work (*'we know not to'*).

- A materially usable 'thing' such as a process, tool or intervention (*'we now have a thing for'*).

- A new practice or way of performing (*'we have a new way to'*).

- Newly heard voices, especially through coproduction (*'we've now heard from'*).

- A new definition or set of parameters (*'we can now specify'*).

Suddenly we're not just locked into mobilising research as a whole (or as an output), but rather considering what parts might be particularly useful for different groups (other researchers, the public, nurses, engineers, charities or whatever). None of this suggests we shouldn't promote outputs, nor should we break down our research into unreasonably disconnected parts, but rather that by focusing on our research as a single entity we're effectively dragging the cupboard round in the hope someone might open it.

And let's not forget we can sometimes mobilise the most invisible part of all this. Ourselves.

> *'The best things I have done for myself in terms of impact have been going on secondment as an academic in external organisations, and seeing how they work, what they are trying to do, and figuring out what I needed to do in those secondments to help them. They were hugely satisfying, but also taught me about how research does and should connect with organisations. It's not about you, it's about the learning you can help to unlock for them. You*

*aren't there to bang your drum, you are there to keep in
time with theirs and, if you are lucky, jam a bit with them.'*
Professor Clare Wood, Professor of Psychology,
Nottingham Trent University

This said, let's also not forget there are some people who aren't
great at going out to speak to other humans without scaring the life
out of them. Probably best to judge that in situ, eh?

WHO PICKS UP THE BATON?

Determining who can pick up the baton of your research to take
it forward is not always obvious. You might know 100% who it
is and have them on speed dial. But you might equally struggle
to identify 'who' because there's no clear audience, your passion
for your topic makes you feel everyone should be interested, or
because you're convinced no one will be interested at all. And
even when it's clear who the stakeholders are, that doesn't mean
they automatically share your energy to continue the race or stay
the full course.

*'Some spaces are dominated by politics, and sometimes
some of the people we want to play with don't want to
play with us. Every situation needs to be judged on its
own merits – if there's things you can do to engage and
make a real difference then do. But if your call is constantly left unanswered, you're probably best to focus
your efforts elsewhere.'*
Lorna Wilson, Co-Director of Research & Innovation
Services, Durham University

The principle of *powering up* can help us get a handle on this
by focusing on how *active* a role different stakeholders can play
in taking work forward. It's little use putting all your energies
into those who aren't interested, and you'll miss opportunities
if you don't mobilise your work to those who really are. As with
all things in impact, it's mildly impossible to convey every possible type of stakeholders, but let's think about this in terms of
a descending scale from active to passive engagement (Table 2).

Table 2. Sliding Scale of Stakeholder Energies.

Those who can *use*	People, groups or organisations who have the power to implement: the research directly, either to benefit themselves or those they represent. This might be a CEO who can use it in their business, a curator who can create an exhibition, an individual who can use it in their daily life, or anyone else who can – without any further approval – crack on and benefit from the research in some way. If you can get the research to them, in a way they can use, you have a fantastic route to impact
Those who can *provide access*	People, groups or organisations who can connect you to those who can use or benefit from your research. Such gatekeepers are the difference between having a pathway that's possible, and a pathway that's blocked. Examples include community leaders (to access parts of society), charities (to access certain representative groups), prison governors (to access prisoners), health service leads (to reach clinical practice), etc. The easiest way to identify them is if you can't reach who/what you need to *without them*. If you can convince them it's sensible to open the door, you have your route to impact. If they aren't convinced of the benefits of unbolting that lock, you could be stuck
Those who can *champion*	People, groups or organisations who champion your work, but are not in a position to use it themselves. They are brilliant for amplifying the messages about your work, sharing your work with others and publicly expressing their endorsement, but if you rely on them for impact, you're relying on their advocacy being *enough* to convince others to change
Those who *listen*	Sometimes people are interested, but that's it. They might feel your work is valuable, important or interesting, but pay no more attention than that. A classic example is someone who follows you on social media, and may like a post, but otherwise stays silent. From a powering up perspective, this type of stakeholder offers goodwill but little energy
Those who *pass by*	Someone who looks to be listening – such as following you on social media – but neither engages nor appears to have any active interest in the messages. The equivalent of having the radio on in the background

Decreasing levels of active use

You'll notice I haven't put beneficiary in this list; that's because it's not an exclusive category. ANYONE or anything can be a beneficiary in some way: the business lead for whom the findings shape their strategy; the local council who now have the evidence by which to make a funding decision; the animals who now receive better care. A beneficiary is simply someone who or something that benefits; there's no rule about where in the chain they have to be.

Breaking it down in this way can help you consider how, where and on who to focus your energies, or at least understand what onwards traction you're likely to get from any given stakeholder. And correctly judging if somehow has the power to use research, or can only shout (however pleasantly) helps you decide whether the baton needs passing in a different direction.

Partner Up ...

Inherent to this thinking about 'who' is about where you fit, and when you need to step away.

It's fairly common for impact models to focus on how the research transacts along the chain, but it's often harder to understand how much of that is on you. Ultimately you can do a hell of a lot to line things up in readiness for impact, but at some point the research crosses the threshold into 'real-world ownership' where it thrives or falls on its own merits. Impact in this way is very much like parenting: just as with children you try to give research the best start, feed it, keep it safe and hope you've done it enough to put it on a good path for (impact) adulthood. You mark progress with milestones (reporting) and pictures on the metaphorical fridge (outputs), at some point wave it off with both joy a tinge of sadness, with a hope of a text now and then as you repocket your much lighter wallet. And just like parenting it can drive you mad, sometimes need a handhold and a cry, and brings a fairly significant need for prosecco.

Implementation is a process of elective uptake by a third party, and so it is not just *our* efforts that matter, but the motivation, autonomy, needs, capacity and authority of stakeholders to adopt research. So when do we dab a hankie to our sobbing eyes, wave goodbye and *transfer power*?

Ok it's all rather less dramatic in practice, but ask yourself two questions:

1. Can I implement my research directly? or

2. Do I need someone else to make impact happen?

For (1), unless the research can't be publicly disclosed, we almost always have the power of communication. We can write about it, present on it, create websites, blogs, videos, performances and all else to get our message across. But if that's not enough to drive change, we can also use opportunities such as training and consultancy (providing expert advice) to help change attitudes, knowledge, strategy or whatever else needs to be addressed.

We can often though only get *so far* via these routes, and need to pass the baton (2) to someone else. For example, we can't change national policy ourselves; we need to get our research in the hands of those who can. Similarly we can't introduce a new drug to the health-care system independently; we need regulators and commissioners to be sufficiently convinced of its merits to add it to care. Closer to home we might find that our efforts to advocate for a change are muted until we engage with louder, better positioned advocacy groups, and it may be as simple as needing our work translated into a different language or format for it to reach who we need it to.

Transferring power doesn't mean giving up, quite the opposite. It means understanding who can take it forward once you can reasonably do no more. The crossover point may happen early in the process – such as MegaCorp International thanking you for your contractually agreed research and running off with it into the sunset to design its new widget – or much later through years of dialogue and trust building with a community. None of us want to hand over our research baby, only to find it's been patted on the head at a meeting but nothing more. But just like our children growing up, we can stay in the research's life, keep an eye on how it's doing and every now and then see if it needs feeding.

'The best impact partnerships run deep. It's more than just people thinking an idea you had was worth mentioning, or providing you with evidence of something they've

implemented. When I think about who I work best with,
it's those relationships where we keep circling back around
each other Columbo-style ('Just one more thing before I
go ...'), asking if there is interest in working together
again or connecting on something else. That might be
involving them in teaching as well as research, or me
inputting into or getting involved in other activities
they are trying to get going. Some of that is sometimes
out of my comfort zone, but they have always been
worth getting into. The more you see of each other, the
stronger the connection and the more your work will
become impactful, as you recognise closer ways of tying
your interests together.'

Professor Clare Wood, Professor of Psychology,
Nottingham Trent University

... But Consider Breaking Up

This is all great, but what should you do if an impact relationship all goes a bit sour? Like on a bad date when his unsolicited views on the appropriate frequency of toilet flushing make the night take a nosedive, or you're ghosted after what you thought was a lovely lasagne fuelled evening. What can you do?

Where possible it's always best to anticipate a break up early, ideally within a formal or contractual process, insomuch as you can set up a kind of impact pre-nup to safeguard your respective interests and outline why or how you might need to decide to call the partnership a day. Obviously you hope not to need it, much in the same way as you hope not to have to avoid texts from toilet-flush-judgement guy, but anticipating the *potential* for a relationship breakdown prompts the discussion about what to do *if it happens*. Whether this is done via handshake or (I'm advised here to say preferably) contracts, this is essentially the act of saying 'we both agree that the research will be used in *this* way, but if it isn't, we need a professional way to say goodbye and go in different directions'. Thankfully research and contract offices are geniuses at protecting research and making sure everyone feels there is a fair process in play. Go see them. And bring sweets, they like that.

'Commercialising intellectual property can be a brilliant way to get a route to impact, but there are risks if the contract process isn't well managed. Some commercial agreements can lock you in with a single company – if they go on to use it that's fantastic, but if they don't you could actually be cutting off your pathways. Over the years I've seen many occasions where contracting just hasn't worked; the company who've licenced the work have gone bust, leading to long drawn out headache-inducing meetings to claw back the IP, or even occasions where a company have licenced the work simply to block others from being able to use it. Thankfully this can be proactively managed in the contract process through things like non-exclusive licensing agreements, meaning you can license it to more than one company and safeguard your routes to impact, or simple clauses protecting academic freedom. I've learned from all these experiences that when it comes to commercialisation, using the contract process to ensure you keep academic freedom and can get back the rights if it goes nowhere is crucial.'

Helen Lau, Associate Director of Knowledge Exchange, Coventry University

Sadly, relationships fail when you don't expect them too. Much like the heartbreaking disappearance of Lasagne Man, things slow down, people stop replying, contacts move on and circumstances change. If you've tried everything to maintain the link, yet are getting nowhere, look for the most decent way to step away, then reactive your impact Tinder and try again.

'At KMb York we help researchers and partners (usually but not always) from the community or non-profit sector. Sometimes the partnership doesn't work out for any number of legitimate reasons. As the relationship broker it is our role to help them work it out or help them break up in a way that doesn't cause reputational damage to either organization or to the individuals involved. This is

difficult work but if the parties can walk away while still respecting each other then that is a successful breakup.'
Dr David Phipps, Assistant VP Research Strategy & Impact, York University Toronto, and Director of Research Impact Canada

WHY? *BE AN ANNOYING TODDLER*

Research doesn't have its own lifeforce. Unless it is activated, kicked along or paid attention to it won't 'do' anything. Much like a teenager in the school holidays. Your research can only power up something which matters to someone, otherwise they probably have other things to be getting on with. As with all aspects of impact, dialogue with those outside academia gives us the answer, and to find out *why* it matters you can delight in becoming an annoying and inquisitive toddler:

'*Why why why why why why why ...*'

have a drink and a nap

'*Why why why why why why why ...*'

The *powering up* principle reminds us to understand not only what research contributes in general terms, but precisely what it meaningfully offers for those who use it.

Helpfully language can do much of the heavy lifting for us here. Language is powerful in impact, not just for finding fancy words for a funding bid or case study but for accurately and meaningfully describing how research contributes. Impact language is the world of verbs, action and change, helping us really understand what traction the research offers for the 'real world'.

If we picture the act of passing the research baton – somehow handing over the learning from our research to those outside of academia – it sets in place the mental image of an onward race. If we can understand, and accurately describe why they'd grab the baton, and why *that* baton, we've worked out what we've powered up.

The question to ask is *what does this research 'do' for them?* See if this list of illustrative words and phrases helps:

What does the research do for the stakeholder?	And that means …?
Enables	Makes possible something that previously wasn't
Empowers	Makes someone more confident, or have more authority, the absence of which was blocking change
Equips	Provides something material, the absence of which was blocking change
Solves	Provides an answer, the absence of which was blocking change
Simplifies	Makes something that complicated easier to use/access/understand
Accelerates	Makes something faster, such as a process, strategic change or achievements of a goal
Amplifies	Makes (e.g. a message) louder or more prominent
Offers a new way to …	Creates an alternative option
Creates the opportunity to	Opens a door or creates a space
Validates	Confirms something, the absence of which was blocking change
Strengthens	Makes a message, position or similar stronger
Improves	Makes (e.g.) the quality or scale of something better
Reduces	Decreases something, such as barriers or waste
Enhances	Makes more powerful, prominent or otherwise higher in quality
Eliminates	Removes something, the presence of which was blocking change
Avoids	Helps stop people going down a wrong path
Consolidates	Draws together or connects things in a way that can now be used

This is not an exhaustive list, nor are the terms mutually exclusive, but they do help bring some focus on what research offers to non-academics, rather than what it does for us.

'When it comes to impact, context and social skills are essential. We must not only consider who to pass the

baton to and how to pass it, but we must understand the nuances of what knowledge/research users want, which is very different to what we think they need. Successful impact requires empathy for the views, experiences and ultimately the wants of those receiving the baton. Afterall, behaviour is driven by what people **want,** *often at the expense of what they really* **need.***'*

Dr Tamika Heiden, Founder and Principal, Research
Impact Academy, Australia

CONSIDER *HOW* THE BATON PASSES...

There is no impact template, because there can't be. Every path is different, so don't think you've missed a memo if your journey looks different to someone else's. At this point it would be easy to jump straight to a list of methods – social media, public engagement, knowledge transfer partnerships and the like – but it's important to first think about what you're trying to achieve.

The route between research and impact can be most simply conceived of as an exercise in *push and pull*. Researchers push out the research, and we need society to pull it. Of course that description is much simpler than it happens in practice, but it reminds us of the need both for our energy to 'get it out there', and making sure it is 'pullable'. We're fairly used to pushing research – disseminating, showcasing, sharing and such like – and there's no doubt we've become far better at more creative and innovative ways to do this in recent years. Blogs, animations, Pint of Science, TED talks and many other such non-traditional formats have not only grown in popularity but have become far more commonplace expectations in funders' expectations of communications.

Pushing research is vital, but for people to 'pull' our research, they need to: (1) know it exists, (2) feel they need it, (3) know how they can use it, then (4) actually use it. That's a fairly comprehensive set of knowledge and actions we're dependent on for our research to be successfully translated. Our improved communications and the open science agenda have done a lot to help people know research exists (1), and whilst we can absolutely do much more, research is at least far more available than it's ever been. Beyond this, steps 2–4 get a little more complicated.

Sometimes there's a ready-made audience for our research, fast tracking ourselves through the 'do they feel they need it' point, but other times we have to work much harder to establish a link in people's minds (or accept we're barking up the wrong tree). I'll just crank up the volume on the 'coproduction and early engagement are vital' broken record here. But we also need to acknowledge that sometimes we don't know who the audience is from the outset, and it is in the act of sharing that we find those who can use it. It is not unusual, for example, for a piece of fundamental science research only to find its non-academic handler when a company who do something we've never even heard of steps forward and says 'aha, I can use this'. Alongside reminding us of the need to package our research in a usable way, visualising the way path from (1) to (4) points us at the further effort needed support it actually being used.

There are, unsurprisingly, no set ways to do this, but the simplest equation is this:

The more you do to make your research relevant for and usable by those who can/should use it

+

The less effort demanded of non-academics to work out how to use it

=

More chance of use

... Before Choosing The Method

By unpacking the various angles of impact, you can more comprehensively map out the journey and thus some of the methods which might be most appropriate. There are already fabulous resources and amazing experts out there who can offer nuanced and detailed explanations of different ways to mobilise knowledge, but I'll keep it simple here and focus on how you might make the decision.

As with everything with impact – just as with research – there is no-one-size-fits-all approach or decisional algorithm that can ever replace human contact. The insights and experiences of those who work or live in the areas we're trying to influence are *the* best guide for the pathway that's going to work. But by way of some examples:

- If your research is of clear benefit to wider society, and/or you need to reach out to a wider audience, media, social media and public engagement can be particularly useful. *A note on social and other media:* No visibility of research is arguably the worst thing for impact, so media is fantastic. Just be mindful though that sharing without being clear on why, or where people can get further information about what to do with it, is the twenty-first century equivalent of shouting at people in the street.

- If you can't reach an audience directly, engage with representative bodies, services, charities, businesses or other agencies. Such groups, who have the interests of their respective groups very much at heart, can be both advocate and guide for reaching those who matter.

- If you are seeking strategy or policy change, engage with commissioners or policy-makers as early as possible to determine how they can use your work, how to package it and what might get in the way.

- If your research isn't readily translatable, work with other disciplines or more applied versions of your own discipline to collectively build something usable.

- If society doesn't yet seem ready for the change, work with those who are already actively vocal about the need, such as charities or civic organisations. Empowering them with research evidence might be the best way to drive change.

- Where possible engage people in the process of research rather than waiting until there's a publication to share. Participatory or collaborative research can bring people far closer to the journey and identify impacts you might never have thought of.

- Be creative and tailor your communications; whether it's a formal report or a contemporary dance, use the method which fits the bill to increase the chances of your research being useful and used.

'Targeting research outputs (either planned or completed) to user groups, and establishing the appropriate methods to engage those user groups is vital yet often overlooked, particularly at the outset of a research project. Whilst there are broad examples of best practice, every project needs to consider questions around research uptake on their own merits, and according to often unique timescales. Success is, of course, never a given, but combined with a clear view of what success might 'look like' a research uptake strategy is an indispensable part of the research impact toolkit.'

Dr Chris Hewson, Faculty Research Impact Manager
(Social Sciences), University of York

PRIORITISING (IF YOU HAVE TO)

Having commented on 'better' impacts in chapter 1, I'm also including a section here on prioritising because – with the best will in the world – there's often nowhere near enough time and resources to pursue all the impacts we want. Sometimes we'll be faced with the need to concentrate our efforts on only a subset of impact possibilities, commonly for assessment but also where resources are tight, so how do you know which to concentrate your *powering up* energy on?

There is no single way to decide which ones to invest in, but let me offer you three guiding questions:

1. What is most meaningful?

2. What has the most reach?

3. What has the most significance?

Meaningfulness is at the top of this list, and if this were a presentation right now there'd be flashing lights and giant arrows pointing at this question glowing in massive screen-filling font size. But it would be disingenuous to suggest this doesn't come under pressure to be displaced by BSEs (Chapter 3) for assessment or other places where scale is key. Thankfully they don't necessarily need to be in conflict. As long as your paths aren't in conflict with what matters – as in a case I saw which accidentally suggested it would be economically advantageous to kill pedestrians – then you're fairly safe to prioritise the paths with the most significance and reach, and head for the most substantial impacts in those directions.

SUMMARY

Approaching impact with a 'powering up' mindset not only expands our thinking about the possibilities for impact, but it recentres our attention on how the different parts of our research can be made useful, for different reasons, and through the energies of different stakeholders. We can't ignore the fact that whilst some of the journey of impact is within our control, some is beyond it. Connections between research and impact are unique, fluid and context dependent; often our task is to keep the impact flame burning as it gets progressively further away from us.

WHAT CAN YOU DO?

Enjoy hunting for places where you can enable, empower, equip or otherwise contribute to making a change, wherever that path goes next. And be as proud of me as I am of myself for not using a Power Rangers reference. Until now.

Ask yourself:

- What aspects of my research can be mobilised?
- To who?

- Why does it matter to them? What does it provide them with the means to do?

- How do I get it to them?

- Who can help me manage the risk of breaking up?

- Am I now thinking about Power Rangers?

Principle 3

THINK DIRECTIONALLY NOT LINEARLY

Flip the problem

'Research branches off in all sorts of different directions, with impacts arising from different aspects of research outcomes and the methods themselves. As a sector we sometimes talk about impact as if it's just the natural result of high quality research, but that assumes a straight-forward path that often isn't there. Change happens in multiple different ways, at different points in time, and through the efforts of many people. We need to be able to understand how to influence change, what that looks like, and how we can demonstrate we've made a difference. By working collaboratively with partners and stakeholders to shape both our research and knowledge exchange, we can fuel how our research creates transformational change. And then – like a squirrel hunting for nuts – know what we're looking for, gather the different stories of effect, and collectively build a sustainable ecosystem.'
Dr Stephanie Maloney, Director of Research and
Enterprise, University of Lincoln

We've talked already about the dominant rhetoric of impact so often being that of linearity, the idea that there is a kind of 'as the crow flies' route across the university–society abyss towards a brighter better future. And we've commented that whilst some research works in this way, it's not by any means *the only way*. Trying to apply a single approach to a very varied thing is, practically speaking, a pretty pointless thing to do. But, equally, approaching it with the equivalent of a blank page and a hopefully poised pen is also unlikely to get us anywhere fast. So if linearity isn't the right foundation, let's shift from conceiving impact as a single journey and think of it more like a courier dropping off multiple parcels. Let's think about impact as journeys with different starting points and different directions of travel.

And yes, this chapter was almost taglined 'hunting for nuts'.

WHY IS THINKING DIRECTIONALLY USEFUL?

If an impact is a change, then there must some way to articulate what the difference looks like.

It's not uncommon for people to use what I'll bluntly call 'non-committal' language when it comes to impact. 'There will be an influence on ...' or 'this had an effect on ...'. Completely understandable – we're commonly taught in many disciplines not to claim causality where it doesn't exist. However, that's not what I'm talking about here. We can usually commit to the direction, just not guarantee the likelihood. For example, as a health psychologist I don't just want to influence people's confidence in using contraception, I want to *increase* it. And someone working on children's literacy is unlikely to want it to get worse.

Impacts can therefore be visualised (see Fig. 4) as a change in the direction or trajectory of something. In some contexts, the aim is to instil an upward change (e.g. increased, faster, strengthened and improved), in others a downwards change (e.g. reduced, slowed and lessened), or otherwise the need to steady things by protecting, preserving or de-risking them. More on that shortly.

FROM 'PROBLEM' TO 'BETTER'

For there to be the possibility of impact, there needs to be the existence of a gap or problem. Otherwise, what are we realistically

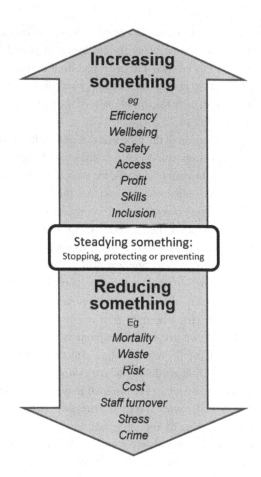

FROM 'PROBLEM' TO 'BETTER'

Fig. 4. Impact as Up, Down or Steady.

trying to make 'better'? And whether this is a global health crisis or something far smaller within your local community, thinking directionally can help you address issues of any magnitude.

Now I'm aware that the word 'problem' is loaded. For some disciplines it is far more natural to express the research starting point in terms of gaps or inadequacies, using words like 'absence of' or 'inadequate' to describe the context. For others however, problematising can feel like we're *judging*, or arrogantly conferring our values on something from afar. Not only that, but what is considered a problem by one discipline may not be viewed very differently by another. None of us should ever feel pulled out of our ethical frame, so to be clear I'm just suggesting that we use the *language* of problems to clarify our starting point, the way research can contribute, and identify the *direction(s)* of the change(s) we are seeking.

Correct characterisation of the problem(s) is one of the most powerful ways to start any impact journey, as it gives you the springboard from which to think about impact goals, for who, at what level, how they can be measured and how you can get there. And *broken record siren* this is superpowered by connecting with people outside of academia. Not sure what the problem is? Speak to people. Not sure what impact goals matter most? Speak to people. Feeling sure but want to double check? You know the drill ....

I'm of course risking horrible levels of oversimplification in this book by splitting 'academia' from 'non-academia' as if we're somehow cap and gown wearing hermits who've never visited a supermarket. Of COURSE we are both; the point is for research and impact we need to remind ourselves that sometimes our vantage point on the world within academia, be it a consequence of access, expertise, detachment or whatever else *can* make the world look slightly different to those who don't have these opportunities (and privileges). It is the danger of assuming, not an absolute likelihood of being wrong, which demands we substantiate the need through whatever conversations, explorations or insights are appropriate.

Back to directions. Impacts are essentially the difference between the situation before the research (let's call that baseline), and that which comes after in some way *because of* the research (let's call those goals). Expressing the difference between baseline and goal is then most simply achieved by using two steps:

1. Describe the baseline as a problem.

2. Describe the impacts as the opposite of this problem.

Step 1: Describe the Baseline – 'What's the Problem'?

Think about your research. What's the impetus for it? What's missing or lacking that you want to tackle? This may or may not be immediately real world facing, for example, it might be a lack of evidence for how theory construct A predicts theory construct B, frustrations that a controversial history is unchallenged, no-one as yet having worked out exactly what shade of blue that very blue thing on a microscope slide is, or the small problem of a pandemic without a vaccine.

Focusing on the reasons your research needs to happen sets a mental (and on-paper) anchor to for what change you could seek to have. If your gap is very academic in nature, then lock into why this problem is an issue for the wider world, however directly any impact might be achieved. A lack of evidence for how theory construct A predicts theory construct B may prevent us designing effective interventions further down the line. An unchallenged controversial history may allow myths to circulate about how current issues should be dealt with. An unshaded blue may limit the diagnostic accuracy of a new tool (with obvious downsides for patients). And it's probably safe to assume the vaccine/pandemic situation speaks for itself.

Whilst there's an endless list of 'gaps' our research could support, thinking about impact as a change in direction (see Fig. 4) we can broadly speaking express baseline problems in one of the four ways:

a) *As an insufficiency*, that is, *too little of something*, for
    example, poor knowledge, little awareness, limited access,
    poor compliance, low efficacy, no policy, lack of skills, non-
    representation, etc.
b) *As an excess*, that is, *too much of something*, for example, too
    much waste, high rates of crime, too much harm, high levels
    of mortality, unnecessary risk, long waiting times, increasing
    (global) temperatures, too much stigma, high rates of burn-
    out, etc.
c) *As a risk if left unaddressed*, that is, *something will deterio-
    rate without intervention*, reflecting circumstances where
    inaction allows something to worsen, such as preserving
    deteriorating historic materials, or stepping in to excavate a
    car park before they dig it up for redevelopment.
d) *As a risk if unvalidated*, that is, *something bad may be
    being allowed to continue*, reflecting circumstances where

something ongoing is assumed, asserted or simply 'accepted' to be ok without any level of necessary check. Examples include untested treatments, fake news, policy decisions without a firm evidence base and any other areas where 'ploughing ahead' without checks risks causing a whole load of damage.

Research can of course have multiple starting problems (e.g. high rates of obesity + poor understanding of nutrition + unproven healthy eating interventions being used to address it). Whilst the practice of impact is always more four dimensional than any way by which it can be written down, these four categories at least provide a mental crib sheet to set out more precisely what the research could address.

Step 2: Describe the Impact Goal(s) – If That's the Problem, What Does Better Look Like?

Once you can describe the problem, you now have a way to visualise what change needs to look like. Impacts are basically the positive counterpart – that is, opposite – of the problem. If we 'flip' the problem we can see the direction of change the research aims to facilitate:

a) If the problem is *an insufficiency*, research supports an *increase*. For example, if the issue is *limited* access, the impact goal is *increased* access.

b) If the problem is *an excess*, research supports a *decrease*. For example, if the issue is *too much* waste, the goal is *reduced* waste.

c) If the problem is *a risk if left unaddressed*, research can *steady or preserve things*. For example, *deteriorating* heritage is *preserved*.

d) If the problem is a *risk if left unvalidated*, research can confirm or disprove something, guiding whether it should continue. For example, *untested* treatments are *stopped*.

Table 3. Characterising Baseline Problems and Impact Goals.

Baseline Condition			Impacts	
Problem Can Be Expressed As:	Example Terms	Direction of Change	Impact Goal Best Expressed As:	Example Impact Terms
A. An insufficiency	Too little Too slow Restricted Limited Low rates Insufficient Unaware Weak Lack of No		An increase	Increased More Faster Widened Higher rates Sufficient Stronger A new
B. An excess	Too much Too fast High rates Excessive Wasted		A reduction	Reduced Lower rates Slower Better paced Sufficient Less Conserved Safer
C. A risk if unaddressed	At risk Unsustainable Unsteady Unaddressed Unsafe		A steadier, more secure or sustainable condition	Sustained Sustainable Maintained Safe
D. A risk if unvalidated	Unchecked Unvalidated		A checked or safeguarded position; or something being continued or stopped	Stopped Ceased Verified Eliminated Avoided Validated Protected

To help you lock this schema into your head, Table 3 summarises the four categories, illustrative words you might use for each (at both baseline and goal), and the shape of change you're really looking to show.

Two final things to say on directional impact:

1. These four categories do not denote the *type* of impact (instrumental, conceptual or capacity building) nor are they related to any specific domain. They are an agnostic shorthand designed to help visualise the relationship between a problem and its associated change, in whatever way that might happen.

2. Research can, joyously and always without apology, have multiple impacts. The reason for pushing for directional over linear thinking is that as soon as we recognise research pathways can spread out in various directions, we start focusing on the changes themselves rather than the singular journey of a piece of research. Some impacts will need short journeys, some long, but at least each path is heading off in the right direction.

SUMMARY

It can be extremely easy to fall into the trap of treating research as a single vehicle which must be put on a path to a single (big shiny) destination. But much like an underground tube map with different lines and different destinations, in practice impact possibilities often splinter off in numerous directions. Turning our internal scriptwriter to directionality can help us accurately articulate the 'problems' at the outset, giving us a clearer line of sight to what 'better' looks like. It can also help us avoid shoehorning expectations of research into unrealistic models of linearity (where linearity isn't possible). Harnessing the language of problems superpowers our thinking and helps us more accurately articulate what changes we can make, demonstrate and be proud of.

WHAT CAN YOU DO?

My advice to you: Shift from thinking linearly to thinking directionally, recognising the power of research to alter the course of things in the real world. Invest time in understanding and describing the problems(s) your research could help address, lock into the language and visualise the shape of change. Knowing what kind of change your research can have makes it easier not only to articulate impact, but also to stay motivated when the pressures of academia kick in.

Ask yourself:

- What problems, gaps or challenges can my research help address?
- How can I accurately describe these 'problem(s)'?
- If they are the problems what does 'better' look like?

Principle 4

EVIDENCE? THINK 'WHAT WOULD JESSICA FLETCHER DO?'

Be on the hunt for clues

Ok this might be a niche title, but if I was ever going to have the chance to intentionally use (shoehorn in) a 'Murder She Wrote' reference, this was it.

This chapter focuses on gathering proof of impact. At the most basic level, any claims of research impact are the answer to the question *What did my research change?* Or more counterfactually *how would the world look different if my research didn't exist?* If you're unsure if you've had impact, remember that if the world looks in any way different because of your research, then you must be on the impact bus.

In Principle 2 we looked at how to power up, and in Principle 3 we looked at how to articulate the direction of those changes. Now we need to think about how to prove any impacts we claim, answering the question *how can we know it's changed?* What should you look for and what evidence counts? And what can you do when the claim you're making doesn't lend itself to a neat and tidy data package of lovely provable proof?

First the elephant in the room – there is no doubt that the need for evidence can be a considerable administrative burden. For the UK's REF process, for example, essentially if it couldn't be proved, it couldn't be claimed. As a result, institutions invested huge amounts of resources to gather corroborating evidence in the form of testimonials, reports and all else, and curate this into auditable records. The extent of evidence gathering required varies across the international sector, but where impact is a feature of research assessment, there remains an incumbency on researchers and research institutions to gather proof of benefits. This can be frustrating, partly because – particularly in resource limited institutions – efforts necessarily shift to gathering evidence, largely prioritising those pathways yielding the most persuasively measurable effects (i.e. will score best).

But let me get off my soapbox, because from a real-world perspective, evidence is needed to avoid assuming there's impact just because the research exists and has received interest. Where researchers are reluctant to gather evidence, at best this comes from a genuine unwillingness to burden people with the labour of providing proof. Whilst much rarer, at worst this comes from a presumption that the research is stupendous enough to automatically and unquestionably shine down on a grateful society, therefore proof is unnecessary. But impact must be corroborated if claims are to pass beyond assumption, in much the same way you'd want your medication to show proof it works and won't turn your head bright green.

For those of you who don't know Murder She Wrote – which is of course fine but we may struggle to be friends – the protagonist is a kindly retired English teacher, who cycles round a cosy part of Maine, regularly (and perhaps suspiciously) stumbling over a juicy murder committed by tampered electrics/a conveniently available vase/poisoned jam. Cue a few conversations with the townsfolk, some helpful pointers for the lovely-but-inept sheriff, and by the end of the show she's revealing the guilty party in front of an assembled set of suspects, dolefully explaining their motive and how a slightly wonky golf club made their guilt undeniable. But I digress.

Why is my love of a detective show relevant? Well crucially Jessica Fletcher is an evidence gathering NINJA. Spotting the clues, asking the right questions and building a picture which unquestionably points to the truth. And this is weirdly analogous to impact. At times it can be confusing what is (and is not) evidence of impact, with the full story only seen when we piece it all together. Sometimes we have to hunt for evidence, sometimes it lands on our lap, and sometimes we have to consider how we can determine the truth in the absence of harder facts. It is the act of detecting and confirming which characterises both impact evidence and Jessica Fletcher herself. And if we're going to take this analogy to its obvious conclusion, I'm hereby declaring my preference to replace all impact evidence submission processes with a gather-them-all-in-the-room-and-unmask-the-real-impact arrangement instead.

HOW DO WE PROVE IMPACT?

When it comes to proving impact, our ultimate aim is to compile evidence – through multiple or single materials – to ensure two things stand up to scrutiny:

- Corroboration that the research contributes to impact (*proof of connection*).

- Corroboration of the impact itself (*proof of effect*).

Ultimately, we are seeking to prove that any impact claim is reasonable and substantiable. Proving the connection is about showing that the impact couldn't have happened, or couldn't have happened in the same way, without the research. There may be irrefutable evidence for this, we may need to compile evidence from multiple sources, or we may have no choice than to try and eliminate all other explanations for the impact we're claiming. And whatever evidence you use, it should be able to stand up to scrutiny by fresh eyes, be they academics, non-academics or Jessica Fletcher herself.

A research mindset can help us here, as we're trying to show 'what causes what', with necessary caveats about how far we can reasonably show causation in the gloriously messy thing called the real world. The simplest, and arguably least frustrating way to think about impact evidence is as analogous to a legal case. Whilst there is absolutely no doubt impact happens even when it's not measured, evidence is needed to push claims *beyond reasonable doubt*. It's about proving that the crime, or in an awkward analogy, the impact, has happened. In terms of the type of information you're looking for – with apologies for a highly technical explanation here – if you're trying to prove something that's numbery, use numbers, but if you're trying to prove something wordy, use words. Instrumental impacts can be well demonstrated with evidence that something is new, changed, replaced or has changed direction. Capacity building impacts are provable with evidence which substantiates the phrase 'now able to'. And whilst conceptual impacts can be harder to measure, evidence is findable where a narrative is somehow changed, for example, topics newly bought into parliamentary debates, the change in the way something is presented in the media, or reports of people feeling less stigmatised.

This book is about principles, but when it comes to something as tangible as evidence, we can't really overlook the practical side. For evidence, principles and practice align when considering *how far you can be certain* of *what impacts*. To that end, below I've outlined four 'levels' of evidence, ranging from hard proof through to claims based on logic alone, with the aim of helping you consider where along that continuum each of your claims may sit. Then in Table 4 there's a summary of some of the methods you could use, and how you might decide which are most appropriate for a given situation.

Hard Proof: There Is No Doubt

Sometimes you'll be in the glorious position of having absolute, hard and irrefutable proof of the contribution and the impact – the smoking gun of the MSW[1] world. Where possible this should be the first point of call for any evidence claim. Some examples of *hard proof* might include:

- Citation in a new or changed policy (showing your research somehow materially contributed to the thing that now exists).

- Sales figures for the widget you designed (showing the company profited by selling the thing that didn't exist before your research).

---

[1]Murder She Wrote, in case you're still not yet as committed to the show as I am

- Clinician feedback that says 'because of this research we were able to change our service ...' (showing that your research was the thing that triggered the change)

... or anything else which categorically links your research to the change being claimed.

Softer Proof: It's Provable When Combined

Often however proving impact is rather more convoluted, especially where claims can't be substantiated by singular or 'neatly contained' pieces of evidence. Just as with hard proof, the requirement is to prove effect and connection, but where there is no smoking gun, you need to compile several points of evidence which when *added together* tell a story of impact beyond reasonable doubt. Typically these types of evidence are weaker, or insufficient *alone* – just as finding a speck of blue paint isn't enough to prove the murderer, until we also then discover Mr Smith painted his house last week ....

Examples of things that can be packaged together can include:

- *Proof of connection*: Information relating to activities you've undertaken which directly arise from your research, such as workshops, outreach events, public engagement, conferences, non-academic articles, performances, consultancy, summer schools, radio interviews or any method you employ. These can answer the question: *what shows there is a connection between our research and the impact we'll go on to claim?*

and

- *Proof of impact*: Information about the change itself, which you can reasonably show extends from the connection for example, via feedback surveys, testimonials, emails, service data, sales figures, service use data, new policies or guidance, revised strategies or anything else showing 'change'. These answer the question: *what impacts arise from these connections?*

Remember softer proof needs connecting, otherwise it's not unreasonable for someone to query if the research and the impact are related at all.

Proxy Measures: It Indicates But Doesn't Prove

Sometimes it is simply still not possible to follow the path all the way from research to proof of impact because you can't quite 'see' the audience. Compare, for example, the ability to assess attitude changes in people who visit an exhibition using exit surveys or interviews, against measuring attitude change in an audience listening to a radio broadcast from home. Irrespective of how many are willing to complete your survey in the first example, they are at least accessible. In the wider media example however it would be extraordinarily difficult, nay impossible, to determine even if they were fully engaged or fixing a tap whilst you were talking. In circumstances where an audience simply can't be accessed, proxy measures can offer an insight into what the impact could be, but stop short of legitimating the actual claim. Examples could include:

- *Engagement figures,* such as audience numbers which gauge how far a message could have reached. These can give a sense of the audience size location (e.g. local vs. international) and nature (such as typical demographics).

- *Gatekeeper accounts:* Where it isn't possible to access the ultimate beneficiaries, consider who *is* in a position to offer comment on their behalf. Examples of this could include:

  o   Teachers, where you can't speak to the pupils directly about the benefit of your new educational method.

  o   Venue managers who can attest to any differences they're seeing in visitor numbers, type or experience.

  o   Environmental leads who can speak about the sustainability benefits to local farms.

  o   Animal welfare officers who can speak about the benefits to pets

  o   Carers, where it isn't appropriate or possible to speak to the vulnerable person themselves

What's brilliant about gatekeeper accounts is they can demonstrate both benefits to the group/thing/issue you're most bothered about AND benefits to the gatekeeper themselves. The teacher who can now use a new approach; the environmental lead who now has a framework to share with other farmers; the carer who feels listened to. They are important links in the chain.

Logical Proof in Uncertainty: We Can Claim If We Eliminate All Other Explanations

If you really can't get conclusive evidence, then you may be left reliant on logic to corroborate the claim beyond reasonable doubt. This is a tactic to deploy *only* if there's no other evidence, and I'd still advise caution; *if* actual evidence could be reasonably expected, then relying on logic alone could trigger a reviewers' siren about the legitimacy of the claim. It's much like a child telling you that 'absolutely honestly cross-my-heart I've done my homework' yet are unable to produce their spelling book when asked.

There are always going to be circumstances where tangible proof isn't available, yet the impact claim is still legitimate. This is typically the case when someone has somehow used, but hasn't credited the research, making it very difficult to establish the link. The lack of credit could be a consequence of anything from naivety to villainy, but irrespective the burden usually shifts to us to try and make the case.

Let's use an intentionally ridiculous example to demonstrate how we could use elimination logic when all other routes are exhausted.

A new policy is issued which says that from now on, a blue cake with three red cherries must be provided at all local government meetings. Hang on a minute, you think to yourself, I published a paper on the productive value of blue cakes with three red cherries for council meetings only recently, and – let's check – yes, this is the ONLY paper published on that topic. It seems very unlikely to be a coincidence, and I can see someone from the local council liked my tweet about the paper. But they haven't credited me! You contact them to confirm if this new rule is because of your research (i.e. form a hard evidence link), but alas no-one replies. Undeterred, you shift to looking for a softer link, such as something mentioning cakes in council minutes, or a comment on social media somewhere. Still nothing. So, with no other way to prove it you deploy Sherlock Holmes logic; 'When you have eliminated all which is impossible, then whatever remains, however improbable, must be the truth'. You consider the facts; before your research there were only biscuits provided at meetings. Your work was published 6 months ago, it is the only published work which proves the merits of blue cakes with three cherries, and now the council have a blue cake rule.

Conclusion: *with no other determinable source by which the council could have been made aware of blue cake benefits, and with it being very unlikely they'd commit funds on a whim, it is reasonable to logically conclude your research was used to prompt this material change.*

Whilst it's not concrete evidence, by eliminating other explanations for the impact, it can be possible to create a logical and credible claim in this way. But be VERY cautious because this ONLY works if you can fully discount all other explanations. Also, I now want a cake.

Identifying Onward Routes; Using Events as Evidence Waypoints

One of the main reasons I talk about powering up (Principle 2) directionality (Principle 3) and is that this gives you a far more expansive map of the points at which you could see change and therefore gather corroborating evidence. And this is particularly important when you can't hang your hat on a single big fat claim. Each baton pass creates an evidential *point of proof*, forming a chain of effects showing how the research progressively makes a difference.

For example, there's almost always an appetite to evaluate knowledge mobilisation activities, such as seminars, workshops and public engagement events. These can be quantitatively or qualitatively evaluated as a thing in and of themselves, giving you a measure of the direct impact they've had on things like knowledge, attitude, awareness, skills or new connections. Evaluation at this point answers the question *what has changed as a result of this event?*

Helpfully these points of proof also give you an opportunity to explore what impact could come further down the line. Whilst future impact is rarely claimable, using evaluation points to identfify the onward direction of impact travel can act as an evidential *waypoint*, lining up what changes may occur (and be monitored) further down the line. What might the people or organisations do differently now? What has the direct benefit (change in knowledge, attitude and whatever) unlocked for them? By asking questions about plans triggered *by* the event, you can answer the question *what further impacts has this made possible?*, giving you a line of sight for future effects and evidence.

WHAT COUNTS AS EVIDENCE OF IMPACT?

I deliberately left evidential formats to the end as it's important to think about what you're trying to find out before you select how you'll do it. You wouldn't pick a research method before deciding on your research questions first would you? Ok don't answer that.

The reason evidence has to be approached in that order is because it can be super easy to get an endorsement on fancy letterhead, but which says precisely nothing about the impact. You have the paperwork but not the proof.

There is no single way to corroborate an impact claims, but by
its very nature any evidence is strongest when it is:

- *Externally voiced*: Created by or expressing feedback from
  those outside academia.

- *Legitimate*: From a credible, believable and trustworthy source.

- *Aligned*: Matching the nature of the claim.

- *Proportionate*: Matching the size, scale or significance of the
  claim.

You're almost always looking for evidence from outside of
academia – from a third party – to push you past the *I swear it
changed, honest guv* level of evidence. There are very valid occasions where you can collect the evidence yourself, such as gathering
feedback after an event, as long as the data itself is somehow *from*
the 'real world'. And remember there may always be need to redact
or otherwise protect sensitive information; just because you need
the evidence doesn't mean people shouldn't expect it to be safely
managed. Without diving into the fullness of data protection principles, be mindful that if your work goes into the public domain,
so might their information. Ask them what they're happy sharing,
and proceed from there.

Anyway, everyone loves a quick reference guide, so let me introduce you to Table 4 (*Table 4, reader, reader, Table 4*). With all normal caveats about this being necessarily generic, this provides a
summary of some of the more common forms of evidence, what
situations they are more (or less) suited to, and how you could
strengthen them in practice.

Table 4. Common Evidence Types and What They're Best Suited To.

Evidence Format	Best Suited to Contexts Where ...	Least Suited to Contexts Where ...	Consider Strengthening This By ...
Surveys	There is a clear, contained or otherwise identifiable cohort of people who have interacted with your research (e.g. workshops)	The audience is unclear, inaccessible or haven't necessarily understood they've interacted with your research	Where possible, survey before and after the interaction to show change (not just satisfaction) Use open questions if you're trying to understand the nature of, or onward path for impact Use closed questions if you want to be able to count or confirm change
Interviews	You are seeking to understand meaning, or issues around adoption/implementation. Especially suited to contexts where you're already in dialogue with the interviewees so don't need to cold call (e.g. local business)	Where you can't reach – or it would place an unreasonable burden/cost on – participants, or when you are seeking comparative metrics	Choosing those with the most authentic and/or credible voice

(Continued)

Table 4. (Continued)

Evidence Format	Best Suited to Contexts Where ...	Least Suited to Contexts Where ...	Consider Strengthening This By ...
Testimonials from third parties	Statements can be gathered from those who can most legitimately make the claim, such as senior organisational leads, the patient who got better, the leader of the council, the museum CEO, etc.	Where there would reasonably be expected to be harder data	Combining with hard data where appropriate, and seeking the most 'credible' voice
	Statements can confirm the contribution and the effect (*because of this research ... this change happened*)	The testifiers aren't credible, or aren't realistically in a position to legitimately make the claim	
Citations in policy	Those glorious occasions where your research has not only been used in the policy-making process, but has been cited	Those slightly less glorious occasions where your research has been used in the policy-making process, but has not been cited	Contact the policy team to request confirmation your research influenced the policy (hard evidence)
			Articulate the means by which your research was fed into the policy-making process, such as submitting evidence to consultations (evidence of connection)
			Clarify – if possible – any particular aspects of the policy which could only have arisen from your research (elimination claim)

Formal organisational reports or data on e.g. visitors/ customers	Materials are independently produced by the third party Both the nature of the impact and the connection to your research are demonstrated in these independent documents The impact you're claiming relates to benefits to these third parties, such as sales, audience numbers, service user experience, etc.	Materials can't be considered independent, for example, where you have created them The materials can't reasonably be considered to show benefits of your research, for example, they show general profits	Get an accompanying testimonial from the third party to show what of the report can be attributed to the research, or to independently confirm they are accurate
Informal communications, such as emails, minutes of meetings	You need to build a bigger picture of contribution/effect, and harder evidence is difficult to find You can reasonably expect references to your research to be included somewhere in these materials	Impacts which are not likely to have arisen through documented channels, or where you research would not have been formally acknowledged	Where such communications are found, link them with other types of evidence to form a convincing picture Confirm if you can use them if they're not in the public domain

(Continued)

Table 4. (Continued)

Evidence Format	Best Suited to Contexts Where ...	Least Suited to Contexts Where ...	Consider Strengthening This By ...
Practice (or similar) guidance change	Your research has been used to change the way something is done, particularly where it is enshrined in formal best practice (e.g. practitioner code of conduct)	Practice changes are not clearly attributable to your research	Get other evidence, such as a testimonial, which expresses how this change wouldn't have been possible without the research Engage with key representatives, such as professional bodies, to influence their wider messaging
'Swift exit' methods, such as dropping a token or voting card in a box	Events where there is a clear audience, but who are too transient for fuller surveys to be viable Where basic metrics are useful markers And/or where more significant forms of evidence aren't possible	The nature of impact as too nuanced to be demonstrable by simple counts	Making sure the question against which participants vote is clear and relates to change
Other people's evaluations	Where others – including other researchers – have evaluated something your research produced (such as an intervention, guide or product), demonstrating its benefits	Where such evaluations have been undertaken, but the benefits remain in the academic or theoretical domain	Follow the evaluation trail; has their evaluation led on to anything further? For example, has a service decided to commission it, or decommission something else?

SUMMARY

Evidence gathering is essentially the task of completing the sentence *because of this research,* _____ *changed, and we know this because*_____.

Corroborating impact is an exercise in piecing together the evidence that legitimately and reasonably pushes claims beyond doubt. We might have the luxury of hard, irrefutable, smoking-gun evidence or we might need to assemble various softer clues which only when pieced together tell the story. And just like a detective show, when all else fails, sometimes all we can do is fall back on pure logic to eliminate any alternative explanations.

What we're ultimately trying to do is show that research contribution powered something up, which led to some level of change outside of academia. For example:

The research enabled us to **simplify** *the process by which people could apply for charitable support, and from that we've seen an* **increase** *in the* **number** *of people applying for, and* **more people being successful** *in receiving these funds.*

Just as with research, think about how you can find the information in the most sensible and realistic way, and be clear on any strengths or limitations that data might have. Where evidence is strong, that's probably enough on its own. When it's not, it'll need combining with other information to legitimately make the case. To think like Jessica Fletcher:

- Look for the clues which show what's happened.
- Talk to those who have their ear to the ground.
- Keep your eyes and ears open.
- Ask sensible questions.
- Assemble your case.

And if you can do that on a bike, whilst cooking clam chowder, smiling sweetly and helping solve the odd murder here or there, all the better.

Be more Jessica. Be an evidence Ninja.

WHAT CAN YOU DO?

Focus on *how* you can know if a change has happened before locking into what material piece of evidence you might be able to grab. That way you'll always know your evidence does the job you need it to. Where possible also prompt proof of connection by making it harder for people *not* to reference you. This might be by creating an identifiable 'thing' (e.g. a set of recommendations, toolkit or learning materials) ideally with a DOI[2], or by being pushy about the need to forge a link (*to use this research, please cite it as …*). Build relationships, keep your eyes open and don't be afraid to smile sweetly (yet with a terrifying sharp detective brain) for the proof you need.

Ask yourself:

- How can I show connection?
- How can I show effect?
- What mix of evidence types could I use?
- How easy have I made it for my work to be properly referenced (and therefore 'trackable')?
- Who have I established links with to 'follow the trail'?
- Have I watched enough MSW? Trick question, you can never watch too much Jessica

---

[2] Digital Object Identifier – a unique string of numbers, letters and symbols assigned to online materials (such as articles and books). For more see https://apastyle.apa.org/learn/faqs/what-is-doi.

Principle 5

CREATE A HEALTHY SPACE

Make impact feel safe and supported

Research environments aren't static. They evolve, they respond, they grow and they adapt. What we need to do is strive for a positive, diverse and inclusive research culture, where everyone is empowered and feels safe to ask questions.

Lorna Wilson, Co-Director of Research & Innovation
Services, Durham University

This chapter is designed to help those of you trying to support or instil a culture of impact, be that within an institution, team or project. If you're not in a position to directly influence this right now, I still urge to you read, or at least revisit this in due course, to set a clear picture for what 'healthy' approaches look like. I'll be using the term 'institution' as a bit of a shorthand, and ask you to read that as organisation-wide, department, centre, school, faculty, team or any other clustering of people you're in. Whether you're happy with them or not.

Before we go any further it's worth reflecting on the fact that typically very few people are able to exert significant force on the environment of the institution as a whole. Universities are complicated beasts, with layers and hierarchies, cultures and subcultures ... you only need to look at the passive aggressive signs in kitchens to see how varied compliance with rules and an unwashed cup can upset the cohesion of an entire department. But institutions are not as yet – unless I'm very much mistaken – sentient, which reminds us that culture is built by people doing all manner of strategic, procedural and hand-wavingly motivational things.

It's also worth saying that every context will be different. Some institutions are already well underway in embedding a positive culture, some haven't yet started, and others are active in impact but not necessarily doing it in a 'healthy' way. Some have lots of cash whilst others only have enough for their bus fare home. It is an incredibly mixed picture, meaning that the language in this chapter by necessity talks about 'risks' and 'opportunities' as only you will know how far along the impact culture journey your institution currently is.

To manage your expectations, this chapter won't focus on highly cited change management techniques or mantras of organisational effectiveness. This is partly because I can't do any level of justice to them compared to the brilliant strategists across the sector, partly because I don't fully understand them, but mostly because we can easily get lost in frameworks rather than stepping back to reflect on what decent behaviour looks like first. This chapter therefore is about being a person within an institution, and thinking about how we can individually and collectively make the environment a more impact-supportive place. Add whatever organisational strategies and frameworks on top as you wish. Go on, fill your boots.

WHY IS THE RESEARCH ENVIRONMENT IMPORTANT?

There is a tendency for the role of the institution to be overlooked in narratives about impact, treated more as an invisible container for the eclectic and individual efforts of researchers. But the environment in which we work can motivate or drain us, and the way is embedded (or not) can frame how impact is understood and supported.

We mentioned in Chapter 3 that institutions can get a bad rap for driving activities to comply with external agendas, even when not complying would have serious consequences for income and organisational sustainability. We also have many examples of institutions declaring genuine appetite to support society in one way or another – the University of Lincoln's 'Permeable University' manifesto being a great example of an institution seeking to be an active partner in social change. It's fair to say the relationship between institutions and impact can be complex.

I toyed with making this chapter one on developing an impact strategy. Which incidentally is a really good idea and you should definitely go make one. But I decided against such a narrow focus because a strategy alone – however needed – is not enough. So many institutional strategies which purport to support impact essentially say 'we're gonna do a crazy load of research, and we've got major ranking aspirations, and there'll be impact obviously because we're brilliant'. But in much the same way as my gym-going is codenamed 'Operation Elle McPherson' – because obviously by writing it down I'll transform myself into a six-foot tall supermodel – without a deeper focus on what is needed to make it happen it is impossible. And in terms of Operation Elle, I will of course be revisiting the earlier section on embracing failure.

> 'As with mission statements, will universities end up with these staple impact strategies – 'one strategy to rule them all and in darkness bind them' – dead before the first letter hits the page, not walking the talk but paying lip-service to their political context and to each other? We also need to find our institutional intrinsic motivation and have impact literacy running through our veins (which is entirely possible, both in research and in teaching). Truly supporting and rewarding impact means breaking down the walls between university services and policy units, integrating impact into HR, Comms, Quality Control, Integrity & Compliance etc. It means stressing the pathway more than the destination, taking care of interactions and relationships more than indicators and rankings.'
>
> Esther De Smet, Senior Policy Advisor, Research
> Department of Ghent University

A Moment on Challenges and Resistance

The reality on the ground is often that we are running around simultaneously delivering on multiple (and sometimes competing) research, teaching and other agendas. Our working lives are commonly not neatly delineated across these areas, and let's all just admit that we catch up on emails about one thing whilst in a meeting about another. Any approach which silos expectations for impact without recognising this can never really mirror the reality of trying to do impact, with the usual result being people pulled in different directions, hating impact and feeling generally burnt out.

The introduction of impact to a research environment can bring a mix of emotions. My experience suggests this ranges from applied researchers welcoming the opportunity to have their work recognised and validated as a legitimate part of academia (having previously felt like the poor relative of more heavy duty theoretical investigative counterparts), through positive acceptance more broadly, through nervousness (including apprehension about added pressures and a sense of 'test anxiety'), to downright anger.

Resistance to impact feels to have diminished – but not disappeared – as the sense of sector wide commitment to social responsibility, familiarity with assessment and the integration of impact into core research practices have each grown. But there are still many people who are at least uneasy about impact, which can make building an impact culture a tricky thing.

I find it's rare to find anyone claiming they simply don't care if their research ever helps anyone or anything, and even rarer for someone to actively hate the prospect. So it doesn't seem to be the idea of societal benefit that's the issue; it's the baggage of expectations, efforts and formalisation that really seem to get people's backs up. It can be easy to read resistance as stubbornness or unwillingness, but more commonly resistance seems to reflect one of four concerns: workload, principle, uncertainty or experience.

Workload pressures take little explaining; academia is routinely characterised by super busy, super stretched work lives and impact often represents 'yet another thing' to add to this already busy mix. This is compounded when impact falls beyond working hours, is unrewarded (e.g. in academic progression), or understanding of how much work is needed is overlooked.

Resistance on principle is almost a pushback on 'what it is to be an academic'. Certainly in the early days of formal assessment in the UK, many of those resisting impact felt that it was unreasonable for academics to be tasked with implementing (and even worse measuring) societal change. This was perhaps even more marked in disciplines further from the 'impact zone' who felt impact was an unrealistic and unachievable expectation of their work.

Resistance through uncertainty more commonly reflects nervousness and apprehension borne from feeling underequipped or underskilled, not understanding what's expected, anxieties related to rule interpretation (including penalties of accidentally breaking rules) and seeking to perfect cases without knowing what 'perfect' looks like. This can be easily compounded by a selected institutional focus on impact *star players*, reinforcing colleagues' views that others have the skills and they do not.

Resistance through experience reflects a drop in willingness through negative experiences. For instance, colleagues finding the case study they've worked extremely hard on isn't being submitted for assessment because someone else's is judged to be 'better'. Or the colleague who has spent many hours building a relationship with a local charity only to be criticised for not producing 'big' enough changes in this time. Or the colleague who was on a pedestal until their impact didn't materialise and they were sent to the subs bench. Managed badly, these types of experience can disengage and disenfranchise colleagues who were previously motivated, but struggle to see what the point is now.

By understanding resistance in these various ways, we can start to see that what looks like obstinance may actually be a degree of practical concern, nervousness, a feeling of one's identity being undermined or a sense of hurt. It could also just be obstinance but let's not go there.

'People can feel scared about getting impact wrong or doing it badly. As individuals we need to be brave, and as institutions we need to make people feel safe to make mistakes.'

Lorna Wilson, Co-Director of Research & Innovation
Services, Durham University

Such challenges can be substantial, but they also present clear opportunities to create a clearer, more inclusive and more rewarding environment for impact, which not only recognises difficulties but makes it safe for us to explore them.

'There should be a safe space for talking about responsibilities in impact. Of course one might argue a researcher has a social responsibility (some might even propose 'accountability') but this argument should be handled with care too. Impact is ideally a group effort. We also do not want impact to feel like this inescapable weight bearing down on all levels of researchers, all struggling with their individual engagement. How can we find ways to infuse the act of doing research with impact without having people buckle under the press of responsibility. In this too policy makers and university leadership have a role to play.'

Esther De Smet, Senior Policy Advisor, Research
Department of Ghent University

Impact shouldn't hurt, so how do we rethink it? Let's start by returning to the principles of impact literacy, this time for the institution.

INSTITUTIONAL IMPACT LITERACY

If individual impact literacy reflects the depth of understanding about research implementation, institutional impact literacy reflects the depth of understanding needed to ensure organisational conditions support impact. Remind yourself of the Impact Literacy model (Fig. 3 in Chapter 2):

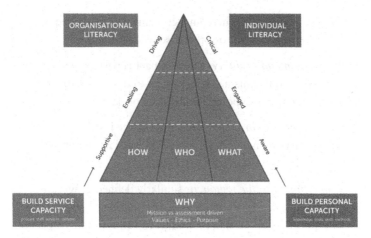

Revised (Extended) Model of Impact Literacy [1]

Until this point in this book we've talked about the right-hand side of the model, focusing on individual literacy and how that can grow through experience and the application of principles. For institutional literacy we now shift to the left and think about the systems and structures needed to build service capacity in the organisation itself.

Just as with individual literacy, institutional literacy has four components:

WHY (The Purpose)

WHY reflects the rationale for embarking on impact, and the values encased in that purpose. However here the query is pointed away from the individual and asks *why does it matter to the institution?* To what extent is impact considered something important, rather than just necessary? And what consequences do they foresee for not doing it? Every institution will have a different mix of answers to these questions, but typical drivers for impact reflect:

- *Formal assessment*, where the institution needs to demonstrate the impact created from its research, usually to inform the allocation of funding.

[1] Bayley, J. E., & Phipps D. (2019). Extending the concept of research impact literacy: levels of literacy, institutional role and ethical considerations [version 2; peer review: 2 approved]. *Emerald Open Research*, 1(14) (https://doi.org/10.35241/emeraldopenres.13140.2)

- *Income generation*, where impact forms part of the requirement for applying for or otherwise seeking funding.

- *Reputation and visibility*, where impact is part of the identity the institution wants/needs to project to the sector, for example, to stand out to potential investors or partners.

- *Commitment to broader change*, where the institution commits to shared 'higher' level initiatives, such as any of the 17 Sustainable Development Goals goals.

- *Commitment to the region*, or similarly defined geographic area, where the institution honours an expressed responsibility to support civic, societal, economic, cultural or other aspect of life in its local context.

An institution may be driven by one, some or all of these. Committing to impact from multiple angles can be brilliant for making impact a priority, but it can also risk conflicting messages if (e.g.) colleagues are encouraged to build local partnerships, yet then only rewarded for international effects. The presence of institutional drivers for impact gives us an amazing catalyst to do good in the world; the trick is not to tie ourselves up in knots doing it.

WHAT (The Policies)

For institutions, the *WHAT* of impact literacy relates to the things in place to 'lock in' their vision and approach to impact. It relates to the tangible, instrumental or material declarations expressing why impact is important. Typical examples include policies, strategies, manifestos the institution signs up to, mission statements and anything else setting the remit for impact within that organisational setting. *WHAT* sends the signal for what is considered to matter, what is supported, what is not and how this fits into the wider institutional mission.

HOW (The Processes)

HOW relates to the practices and processes in place to support the translation of research into practice and the compilation of any evidence. This includes support for engagement (with businesses, policy-makers, cultural establishments, the wider public or any other group), communications, impact planning and delivery, evidence gathering, capacity building activities and any other practical ways in which impact can be facilitated. Their existence (or not) practically underpins the achievement of impact, but also signals to the wider institution what value is placed on impact as an activity.

WHO (The People)

Within *institutional* literacy, WHO relates to the people whose efforts and skills support impact within the institution. Impact is a team sport, including those who produce the knowledge (researchers) and those who facilitate, manage, lead, communicate, connect and evaluate its benefits outside the university walls. Whilst many of these skills are dispersed around the institution across different people's roles, it's increasingly common, and largely the default structure in the UK at least, to have designated impact officers and more senior leadership focused precisely on this area of research practice. Some institutions have a single named individual to 'do' impact, others have large dedicated teams spread across both central services and faculties. But there doesn't seem to be a single best structure, size or way in which impact functions are dispersed across an institution, other than the obvious dangers of relying on a single 'lone wolf' to carry the weight of impact possibilities.

> 'For many years in the UK impact was characterised by 'lone wolves', single impact officers tasked with not only making impact happen, but ensuring there was enough robust evidence for assessment. Unsurprisingly these people got burned out, as the enormity of impact could never realistically be managed by one person alone. Thankfully there's now a large community of practice, and although we still have some people carrying the weight by themselves, impact is seen as a collective responsibility far more these days.'
>
> Dr Chris Hewson, Faculty Research Impact Manager
> (Social Sciences), University of York

To be impact literate is to centre your thinking on real world change; to be an impact literate institution is to centre your thinking on how the actions of creating this change can be best supported. And for impact to thrive, people need to thrive, and that takes both long-term thinking and recognition that everyone plays a part.

> 'Like walking into Mordor, you don't just introduce knowledge brokers into your organisation and have them run free without providing them a safe harbour. Being a knowledge broker in academia is not an easy job. We have a responsibility to not make it even harder. Offer them a supportive environment by investing in sustainable careers (specific progression model and tailormade assessment, long-term contracts), in clear and realistic expectations, in training and coaching (balancing between trust and guidance instead of either micromanaging them or leaving them up to their own devices – pick your favourite type of PI ...), in embedding them into the fabric of the university through collaboration (don't just appoint Impact Champions but create a true community of practice). And sure, by having some funds available that will actually allow them to do stuff ....'
>
> Esther De Smet, Senior Policy Advisor, Research
> Department of Ghent University

INSTITUTIONAL RISKS OF TAKING A NON-LITERATE APPROACH

It is of course entirely possible for institutions to embark on impact without taking an impact literate approach, but this brings a number of risks:

- Considering HOW + WHO, without WHAT risks setting people and processes on a path to impact without clarity on what's expected.

- Considering WHAT+ HOW, without WHO risks creating a system for impact without the necessary skills or culture building activities.

- Considering WHAT + WHO, without HOW risks establishing expectations and skills without the necessary infrastruture to sustain practice.

- Considering any of these without clarity on WHY means that even if the processes, people and policies are good, they might not align with what matters to the institution.

In practice these risks can lead to misaligned efforts, under-delivery on targets, staff frustrations, interpersonal tensions and a suite of efforts which don't bring real benefit to the world outside. And the more room we make for impact to be positioned as something 'extra', the less chance we have of embedding it in a healthy way.

'Mind the gap between individual and institutional impact'. There is a risk that impact is seen as a separate part of the research process. But impact is always an integral part of a multidimensional context related to quality assurance (compliance to standards), responsibility (principles for integrity, diversity, inclusion and equity) and sustainability (as represented by UN SDG's). My vision is that Impact as part of sustainability will dominate all other types of impact throughout the sector towards 2030. So it's essential that sustainability is incorporated at all levels of our impact praxis, both micro and institutional. We need to acknowledge that 90% of what we do towards impact is incremental and transactional and anecdotal in nature, focussing on what is impacted in the short term, also lacking normalised metrics. This needs to be complemented With a more long-term institutional impact approach, building long-term narratives for impact, underpinned by general agreed performance indicators. For this, both individuals as institutions will have to develop new professional skills, so we will become able to demonstrate

impact leadership, driving impact more strategically and intentionally rather than reactively and incrementally.'
Wilfred Mijnhardt, Policy Director, Rotterdam School of
Management, Erasmus University

LEVELS OF INSTITUTIONAL LITERACY

Just as with individual literacy, institutional literacy can grow, and systems can become more advanced.

Table 5 outlines how institutional literacy might present itself from basic support for impact, through to more actively enabling impact, up to more advanced approaches which drive the impact agenda as well as delivering on it.

Table 5.   Levels of Institutional Impact Literacy.

Literacy Level	Description of Level
Supportive (Basic)	Institution recognises researchers must participate in impact-related activities (e.g. impact strategies in grant applications and impact assessment exercises) but has not developed institutional plans/strategies to actively develop impact literacy. Institution supports efforts of researchers but is not actively maximising the creation and reporting of impacts
Enabling (Intermediate)	Institution has developed some policies/plans and is investing in efforts to enable researchers to create and report impact. Institutional policies strongly reflect external agendas, but institutions are not yet critically appraising external models and adapting to institutional context
Driving (Advanced)	Institutions have policies and strategies, are investing in these strategies and in personnel, and have established a cycle of critical stakeholder engagement to drive the ongoing development of impact services

Source: Extending the Concept of Research Impact Literacy. [2]

[2] Bayley, J. E., & Phipps D. (2019). Extending the concept of research impact literacy: levels of literacy, institutional role and ethical considerations [version 2; peer review: 2 approved]. Emerald Open Research, 1(14) (https://doi.org/ 10.35241/emeraldopenres.13140.2)

Institutions displaying more basic levels of literacy tend to recognise the need for impact, but don't have significant or strategic support in place to support this. Institutions at a more intermediate level enable impact through investment or activities to help build capacity or connect with society, but may do so bluntly and without consideration of context or unintended consequence. Institutions at a more advanced level drive the impact agenda, structurally and developmentally investing in impact with deep connections externally.

INSTITUTIONAL HEALTH

'Creating a healthy institutional culture is vital for impact. Not only must that involve researchers, but also teaching colleagues, practitioners, research managers, professional services, students and senior leadership. It is through our collective efforts, and in our support of professional development across all these areas, that we can make the most difference to society.'

Dr Stephanie Maloney, Director of Research and
Enterprise, University of Lincoln

If impact literacy is the understanding of impact, institutional health reflects the extent to which impact is positively embedded in institutional practices to deliver it. There is no one, single or prescriptive way to best support impact. This will vary by institutional type, resources, mission, goals and all else. However, a culture can be considered healthy if it:

- Creates, values and supports the space needed to drive impact.

- Acknowledges the effort needed to deliver impact.

- Acknowledges and values different paths and disciplinary differences between research and impact.

- Invests in capacity/skills across the institution.

- Coordinates internal teams and resources in support of impact.

- Ensures everyone is clear on their roles and how these align to an overarching strategy.

- Builds strong connections with external stakeholders.

- Embeds impact as 'business as usual' from the start, with aligned strategy and processes.

- Delivers impact by developing confident, impact literate staff driving a positive impact culture.

- Locates impact within part of the wider academia landscape.

- Continually learns from best practice.

- Has fair expectation of staff, processes and actual impact.

Healthier environments are like rocket fuel for impact. They are active, reflective and supportive cultures, in which people know how they fit and how impact fits them. It's therefore unsurprising that an unhealthy culture:

- Compartmentalises impact, focusing only on cycles of assessment, grants or specific researchers.

- Delivery treated as the responsibility of one person, team or area of provision.

- Creates no space to build impact into the research process.

- Expects – but does not invest – in staff capabilities.

- Leaves key areas of the organisation disconnected and unaligned.

- Has no overarching strategic vision to guide delivery.

- Leaves staff reluctant, unclear or unconfident about their role in delivering impact.

- Has few or superficial connections to external stakeholders.

- Leaves impact to be considered only at the endpoint of the research process.

- Has a negative or non-existent impact culture, or leaves impact to be treated as compliance with assessment/external mandates only.

- Has unrealistic expectations for any and all aspects of impact.

Unhealthy cultures can manifest in a number of ways. A common marker is a short-term attitude to impact, focused on project-by-project or within assessment timescales only. Institutions can also prioritise metrics and systems to 'fix' their problems with impact (viewing it as a problem with the documentation of case studies), whilst offering minimal support for staff who become burned out trying to deliver more and more social change. This said, it's important to note that unhealthy environments aren't automatically a sign of apathy towards impact; they can also easily be the product of under-resourcing, low staff capacity, impact in the early days of its introduction or other practical factors. But with clarity on what 'healthier' looks like, even with limited resources institutions can seek to address tangible things which can start shifting us towards a more positive culture.

The 5Cs of Institutional Impact Health

To help institutions embed a healthy approach, David Phipps and I developed the 'Five Cs' of institutional impact health.[3] Each category reflects a key aspect of structure or oversight by which can help build and wrap research in healthier practices for impact[4]:

COMMITMENT: The extent to which the organisation is committed to impact through strategy, systems, staff development and operational ways to integrate impact into research processes. It's demonstrated by the presence of strategy, investment (financial or time), support, and training and development, and is vital for shifting beyond platitudes.

CONNECTIVITY: The extent to which teams or areas within the institution work together. This includes not only how they practically connect, but how cohesive they are in de-

---

[3] Bayley, J., & Phipps, D. (2019). Extending the concept of research impact literacy: Levels of literacy, institutional role and ethical considerations [version 1; peer review: 2 approved]. *Emerald Open Research*, *1*, 14. https://doi.org/10.12688/emeraldopenres.13140.1

[4] Available as an Institutional Healthcheck workbook at: https://www.emeraldgrouppublishing.com/about/our-stance/our-impact, or as a fuller interactive tool via Emerald's 'Impact Services': https://impactservices.emerald.com/.

livering on impact agendas. Research institutions are made up of many people and many teams – researchers, leaders, impact specialists, communicators, research managers, information managers and many more – and delivery on any agenda needs collaboration and coordination between these.

COMPETENCIES: The impact-related skills and expertise within the institution, development of those skills across individuals and teams, and value placed on impact-related specialisms. Impact requires effort and skills in brokering research beyond academia.

COPRODUCTION: The extent of, and quality of, engagement with non-academics for to generate impactful research and meaningful effects. Collaboration across the research lifecycle helps determine appropriate research questions, root activities in 'what matters', highlight and address assumptions, frame communications and improve the chance of uptake and impact.

CLARITY: How clearly staff within the institution understand impact, their role in delivering impact, and how impact extends beyond traditional expectations of academic research. Strategic commitment is essential for impact, but for it to be truly embedded in practice, staff need to be clear not only on what impact is, but also how their role supports it. Unless high-level agendas are translated into clear and actionable messages, individuals may feel disconnected from impact, and research unaligned from strategy. Institutions must therefore communicate clearly what impact is (and isn't), the institutional vision and expectations and be transparent on formal requirements which need to be met. But they should also ensure communications are clear on the variability of impact across disciplines.

Whilst it can be difficult to see the wood for the trees when you're in the middle of a burgeoning (or spiralling) impact culture, it can be helpful to visualise the contrast between a healthy and unhealthy version on each of these C's. Put side by side, the difference, and thus the opportunity for improvement, becomes really rather clear (Table 6).

Table 6.   Comparison of Healthy Versus Unhealthy Practices.

	Healthy	Unhealthy
Commitment	Institution has an impact strategy, with dedicated leadership, support and resourcing. Impact is embedded across the research lifecycle built appropriately into workloads, and fairly tied to career development and progression	Institution has no strategic direction, with no/minimal leadership, support or resourcing. Impact receives attention periodically in response to external agendas only, with effort unrecognised in workload planning and progression
Connectivity	Impact-related practices, personnel and goals are connected across the institution, with responsibility for impact shared across academic and non-academic staff. The activities of different people and teams are aligned, coordinated and delivered cohesively	Impact-related practices, personnel and goals are disconnected across the institution. Impact is the responsibility of a single role or components of multiple but disconnected roles, and the activities of those who can support impact are not aligned
Competencies	Staff have the skills to deliver impact, with specialised expertise accessible where needed. The institution supports and invests in skills development for both academic and non-academic staff, providing training and development opportunities	Staff are underskilled and under-confident to deliver impact. There is no or limited investment in skills development for academic or non-academic staff, with little opportunity for training and development
Coproduction	The institution supports multiple and meaningful connections to external stakeholders, establishing partnerships to inform, guide, contribute to and use research. Stakeholder input is embedded strategically into projects and programmes, with end-user needs integrated into research plans	Stakeholder relationships across the institution are superficial and tokenistic. Partnerships may be transient or short term to solve a particular research-led problem, rather than reflecting stakeholder needs. Institution lacks a coherent strategy for stakeholder engagement

(Continued)

Table 6. *(Continued)*

	Healthy	Unhealthy
Clarity	Staff understand what impact is, and how their role is connected to impact delivery. Institutional vision is unambiguous, with staff clear on formal internal and external requirements. Cross-disciplinary differences in impact are recognised, with strategic goals appropriately contextualised	Staff do not understand impact, or how their role aligns. Institutional messaging is unclear, and there is a 'one-size-fits-all' expectation of how impact is delivered across subject areas

EMBEDDING AN IMPACT CULTURE

'*Impact requires that you decentre yourself and your research. It requires the academy, as an institution, to decentre itself. It's about putting audiences, beneficiaries, policymakers, service users, patients, customers, businesses, and communities at the forefront your mind and thinking about what they need and how your research (and the research of others) can support their journey. Impact isn't about leading from the front. Impact requires shared leadership, collaboration, participation. It requires different forms of knowledge and experience to be valued, included, and embedded within research and research design. It fundamentally asks us to reconceptualise how research is conceived and undertaken.*'

Dr Kieran Fenby-Hulse, Principal Lecturer (Staffing and Resource), Department of Leadership, Management and HRM, Teesside University

The impact agenda provides both opportunities and challenges for us to align institutional efforts, resources and skills. Ensuring there is clear understanding (literacy) with a supportive (healthy) environment is vital for everyone to feel like they have a place. We've already said that there is no single or template approach to impact, which makes it easy to assume impact is either a blank canvas devoid of structure, or that impact is only possible by milking the effects of multiple single projects. Actually the best answer lies somewhere in the middle, creating a supportive structure in which individuals and individual projects can thrive. By treating impact more as a programme of activity, we can be strategic without being prescriptive, and help people see how and where they fit.

> 'As the oft used quote goes 'teamwork makes the dream work', and this is particularly true when universities attempt to foster a healthy impact culture. The signal to noise ratio is already high in the field of HE knowledge exchange. If structures are misaligned, or different messages around what is valued cross-cut each other, academics will rightly disengage. That is not to say that good practice comes in easily digestible forms, impact teams must work together to establish what works for their institution, and the constituent parts thereof.'
>
> Dr Chris Hewson, Faculty Research Impact Manager
> (Social Sciences), University of York

And it's here we can use the 5Cs to support our thinking. If you want to start the process of embedding a healthy culture, think about these things you can do:

Show there's commitment by having:

- An impact strategy.
- Dedicated support and advice available for impact.
- Clear impact leadership.
- Support throughout the research process form planning through assessment.
- Dedicated systems to support impact information.
- Impact development opportunities for both academic and non-academic staff.

- Impact development opportunities for students.
- Incentive and reward structures which recognise (and fairly review) impact-related work.
- Impact built realistically into workloads.
- Sufficient resources (internal or external) to support impact delivery.

Whilst it might not be in your power to create an impact strategy or establish leadership, are there ways you can make your groups' approach to impact more strategic? Or find people to lead the way? If you can't establish material rewards, are there ways that people can be made to know their efforts count? Commitment is not about high-level declarations alone, but also about finding ways to support, acknowledge and build capacity in impact however that is realistically possible.

Connect people and teams by ensuring:

- Teams within the organisation who support impact know about each other.
- Teams within the organisation who support impact work together.
- Teams within the organisation work cohesively (i.e. work well together and towards the same aims).
- The activities of teams/departments are aligned with the overall strategy.
- Everyone who needs to be is included in impact provision.
- Impact activities are coordinated.
- Different disciplines are facilitated to connect.
- Skills and expertise are known about, shared and valued.

Academia makes a difference in so many ways, even if we only attach formal names (like 'impact') to some of them. Recognising the various and fantastic ways we can influence the world is the first step to harnessing this collective power for change and embedding impact as part of the wider environment. By understanding how impact sits alongside other aspects of academic delivery, not only can we ensure we honour these endeavours too, but we also visualise how to use these as opportunities to 'power up' our impact (as per Principle 2). Fig. 5 illustrates some of the key activities we do within academia (outer circle), what this offers for the opportunity to make a difference (middle circle), and example terms for what this kind of making a difference is sometimes called. I'm always minded to think like this, partly to make it clear that impact is one of the ways (rather than *the only way* we make a difference), partly to ensure we don't unintentionally disregard the difference-making efforts of those in a different category, and particularly to see at a

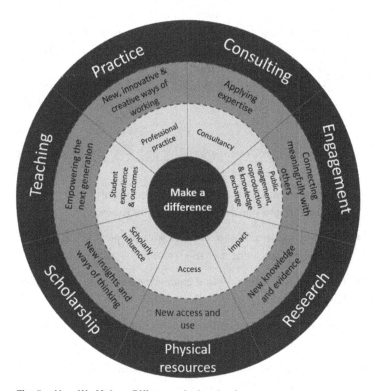

Fig. 5.   How We Make a Difference in Academia.

glance the myriad of opportunities we have to amplify our impact by using the activities we already do.

And in all this, we need to absolutely, completely and with explicit recognition of impact as a team sport, connect academics, knowledge exchange experts, research management specialists, communications gurus and everyone else who wears an impact bib. Not just when the bid is due, or when a case study needs some cheerleading, but throughout the research process and across everything we do.

> 'Work with, rather than around your Research or KE office. You might feel like they're trying to restrict you, but the opposite is true. Colleagues in these teams not only have a clear eye on opportunities to translate research and contractual ways to protect your academic freedom, but can also help make sure you're speaking in a language the outside world can understand. 'Impact' often doesn't mean anything to businesses, but knowing what they do want to achieve and how research can help them gives you a powerful route to change.'
>
> Helen Lau, Associate Director of Knowledge Exchange,
> Coventry University

Support coproduction by:

- Supporting engagement outside of the institution with both commercial and non-commercial organisations.

- Creating opportunities to connect with people and organisations who might inform or use research.

- Ensuring there's support and guidance on coproduced and collaborative research.

- Making research externally visible from the start to encourage collaborative opportunities.

- Where necessary, providing formal or contractual support for partnerships.

- Establishing and encouraging involvement with networks outside the institution.

- Recognising the importance of, and facilitating, trust-building activities.

- Supporting researchers to connect *out* to society, and non-academics to come *in*.

Everything about impact is supercharged by coproduction, but a positive attitude to it isn't enough. Institutions need to put processes in place, create the opportunities and honour the time to build trusted relationships to really embed coproduction as part of the research process.

> *'Good impact is built on relationships. Relationships take time, effort, tea, sticky buns, and meetings where you both get giddy over the same ideas, are moved by the same challenges, and fundamentally want to move in the same direction and support each other in doing that. This goes for relationships with charities, colleagues, commercial partners and yes, in an ideal world, the senior management of the university. If you can get your Dean, PVC or VC to care about the things you care about and understand the challenges, then that's where things can start to connect across your university and resources can follow. But never underestimate the power of tea, cake and honest conversations.'*
>
> Professor Clare Wood, Professor of Psychology,
> Nottingham Trent University

Support skills by ensuring:

- Researchers have the skills to plan, create and monitor impact.

- Research managers have the skills to support impact.

- Leaders have the skills to oversee and coordinate impact.

- There is expert advice available for impact.

- The institution recognises and invests in development of impact-related skills.

- There is training available to build impact skills.

- Training is available to all, not just targeted researchers.

- There is specialised advice available for intellectual property, contracts and legal issues.

- Skills are shared between teams.

When it comes to skills, it's important to recognise there's no reasonable way a single individual could be expected to have them all.

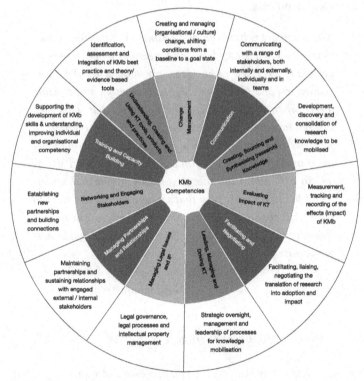

Source: Adapted from Bayley et al. (2018).

Fig. 6.   Knowledge Mobilisation and Impact Competencies.[5]

---

[5] Bayley, J. E., Phipps, D., Batac, M., & Stevens, E. (2018). Development of a framework for knowledge mobilisation and impact competencies. *Evidence and Policy*, 14(4), 725–738.

The breadth of skills needed across impact – summarised across 11 categories in Fig. 6 – can only realistically be held across multiple people. By understanding what skills are needed, and where they can currently be found, leaders can harness the capabilities already in place and identify the skills gaps to be developed. The task therefore is to locate expertise across the institution, connect them where they exist, build capabilities where there are gaps and develop people's skills to establish a sustainable and healthy approach.

Ensure there is clarity:

- On what impact is.

- On what formal agendas drive impact.

- On the institutional vision/mission for impact.

- On how different colleagues contribute to impact.

- On how impact-related activities can be acknowledged in appraisals and progression.

- On language which helps people feel included (e.g. research rather than science).

- That there is no one-size-fits-all version of impact.

- That impact sits beyond traditional markers of academia (e.g. citations).

A clear sense of what impact is, why it matters and how people contribute is vital for people to engage. You can have all the commitment, connection, coproduction and competencies in the world, but if there's myths, mists and misunderstanding, things will start to derail pretty quickly. Clarity ensures that combined efforts for impact are aligned, and any necessary culture shift brings everyone along for the ride.

> 'A recognition that pathways to impact are diverse is also a recognition that there are significant issues of equality and fairness at play, that need to be carefully considered. Institutions should build a picture of the many different

*forms of potential impact across their research portfolios, as
they seek to build an equitable and achievable support structure. On a managerial level, operating an 'impact pipeline' in
this way also ensures that opportunities for impact are supported in a timely manner – a prerequisite in many fast mov-
ing areas of research engagement and innovation.'*

Dr Chris Hewson, Faculty Research Impact Manager
(Social Sciences), University of York

SUMMARY

The aim of this section has been to shine a light on how to wrap a
protective blanket around impact through purpose, policy, process
and people. Introducing and embedding impact into an institution,
or part of an institution takes work. Every circumstance and starting point will be different, so quite how you go about it will depend
on the context in which you're working.

Colleagues are rarely unaware of impact before it's institution-
ally fanfared, but care is still needed to communicate clear expectations, reset or refresh understanding, align practices and address
myths borne from unease, experience or exhaustion. Perhaps the
most basic strategy to adopt is 'engage not enrage', focusing on
building a community collectively striving to make a difference,
rather than fuelling one which competes itself into heartache.

The thing we can't do is *nothing*. By locking your line of sight to
the healthiest picture of an impact environment, you can positively
and progressively reshape how it's supported. I said at the start
this wasn't a book full of change theories and management frameworks, but they are extremely useful if you're needing to drive
change at a more strategic level. Seek out not only the theories but
also the people who know about them, because weirdly enough
there's a fair few experts knocking around the university sector
who know this stuff pretty darn well.

Whether you try and change one thing or 20, it's about getting
traction towards a more supportive, collective and healthy situation. And for those of you so inclined by the collective reference,
resistance is indeed futile.

WHAT CAN YOU DO?

This chapter is awash with 'things you can do', so I won't reiterate them here. Instead I'll just say this. When trying to embed a culture of impact, I urge you to shift away from attitudes which:

- Ignore the competing pressures on individuals.

- Treat impact as a natural consequence of good research.

- Seek to just milk more impact out of individual research projects.

- Misread or assume impact apprehension as stubborn unwillingness.

- Overlook entrenched academics traditions.

- Ignore the glorious mix of skills and opportunities we already have within the research community.

Ask yourself:

- What drives impact at my institution?

- How well is impact understood in my institution?

- Is the institution clear on expectations for impact?

- Am I clear on how I contribute to impact? And how others contribute?

- Whose skills could I make use of? Who could make use of my skills?

- What is in place to help me collaborate with those outside of academia?

- What attitudes or apprehensions need addressing?

- What – across the 5Cs – could we start putting in place to embed impact in a healthy way?

Principle 6

OWN YOUR EXPERTISE BUT DON'T BE A JERK

A tiny nugget of gold in a whole bucket of nuggets

'One of the things I've worried about when working on impact is the reaction of academic colleagues to the way an organisation positions me as an 'expert'. For many of us who have been through the bruising process of peer review, we are very sensitive to the idea that it is somehow inappropriate or unprofessional to claim to know about how things work in practice outside our world of academic complexity and caveats. But remember your impact partners value your externality, your access to theories and findings that they may not have. Your general academic knowledge about the topic (that you take for granted) is a source of valuable insight and new ideas for understanding their problem and your expertise adds to a wider conversation about what they can do. Often your work is one tiny nugget of gold in a whole bucket of nuggets that you and others offer.'

Professor Clare Wood, Professor of Psychology,
Nottingham Trent University

I remember once a doctor arguing with me that I'd had a baby. I was 18 and most definitely had not given birth, not that he'd believe me. I'd gone into an appointment for a damaged shoulder and somewhere between the door and the chair had apparently conceived, gestated and sprung loose a now absent mini-human.

As he continued to try and convince me to just *admit* it, it baffled me that he either considered me a liar, or thought me so forgetful I'd just overlooked the whole event and would suddenly recall popping one out the previous summer between shifts at Marks and Spencer. The thing is, he was SO adamant that his knowledge outstripped mine and that he must be right. And even when I declared a clear no, I can only assume that his perception of the power differential led him to conclude that it MUST be me that was incorrect. Because it was in biro in the notes. And anyone knows if it's in biro it's unequivocally correct.

Lesson learned? *The assumption of being right, particularly when in a position of power, could be oh so very wrong.* To ask if I'd been pregnant would be fine, but to argue with me that I had is downright daft.

Yes it's a flippant chapter title, but I can't find a better way of saying it. Well I can but apparently those particular words 'aren't appropriate'. This principle, which could be (but isn't) more maturely worded as 'contribute but don't dominate', relates to the need to recognise the value you bring, whilst also recognising the importance of valuing the expertise of others.

I'm not presuming you or anyone else is a jerk, but it's also fair to say that the pressures to 'do' impact can unintentionally (or intentionally, depending on your personal life choices) open the door to arrogance. This can manifest in a number of ways, but perhaps most commonly in presuming to know what's needed, rather than listening to the people whose lives are affected. Not only can that take us down entirely the wrong path, it can reinforce or widen – rather than reduce – gaps between academia and society. In contrast, some people are so far from being jerks they feel they can't own any expertise simply because others in the room seem to have more.

There are in essence three broad categories of confidence in academia: under, over and just about right. Let's call this the Goldilocks Paradigm to make it sound fancy.

At one end of the scale we have under-confidence, the cold porridge of this Goldilocks-impact analogy, and something which is far more common than is fair. You may already be very familiar with the term 'Imposter Syndrome', but if you're not, this is the underlying and fairly relentless sense that you're a fraud, and you're really just biding your time until someone finds you out and denounces your professional fakery. In academia it's fuelled by sector messages that only fancy ('excellent') research marks you out as an academic, and that only 'big' impact counts. In impact it presents itself as a reluctance to drive research forward because *who am I to tell people what they should do?* The thing is, so often research doesn't play out as planned – small sample size, non-significant findings, failed bids, etc. – and impact is only partially under our control within academia anyway. Far too often we internalise these as personal failures, a sign of not being good enough. *Spoiler: that's rubbish.* If you don't recognise yourself in these points, fabulous, you've avoided one of the most corrosive ways in which academia operates. But if you do, then I hope this book acts as a way to not only find yourself in impact, but to also say 'screw that' to imposter syndrome and recognise that those frustrating experiences of research are actually just your immersion in this messy world of exploration.

At the other end of the scale we have overconfidence, the burningly hot porridge of attitude which can too easily scald those who come into contact with it. An arrogant position dismissive of other approaches, most routinely typified by the conference delegate who steps up with *it's not a question, more of a comment* then proceeds to wax lyrical about their own work until you're all late for coffee. In impact, this is most easily characterised by assumptions such as *my research is bound to have huge impact* or *I already know what people need, even if they don't know yet,* with obvious implications for reinforcing public attitudes towards ivory-tower snobbery. I remember a few years ago being asked to speak with a researcher about a bid because, despite it being a funder requirement and with a looming deadline, they hadn't yet engaged any public contributors. When I pushed, they refused, advising me they wouldn't be sharing the work until it was *perfect.* With even more eye rolling they explained that public involvement was unnecessary anyway as of course it was important they'd been *working on it for four years.* Sadly we didn't meet again, and as time spent writing is obviously a direct proxy for what the public wants, I can only console myself for missing the inevitable wailings of public joy as the project was finally unveiled.

Jerkness can of course be accidental. Within academia we are taught to (or taught to aspire to) become self-sufficient leaders and experts in our field. Nothing wrong with that, that's how knowledge and disciplines grow. But in impact, positioning ourselves as *the* expert risks a sense of superiority, disempowering those outside of academia and treating their wisdom as inferior. And we can also be jerks to colleagues by trivialising their impact-related successes or leaving them out of conversations. All generally bad moves.

Sadly jerkness can also be intentional. We can all bring to mind examples where arrogance extends to catastrophically inexcusable behaviours such as bullying, plagiarism, data falsification or other completely sucky things. I've seen too many times colleagues convinced they have no value because others have told them their work doesn't have impact, and seen people erased from the history of case studies to show a 'neat' story. Given its connection to the 'real world', of all the areas of academia, impact should be a far more equal and democratised conversation, wherever in the academic hierarchy you sit. But impact is part of the research sector, which means it isn't immune to some of poor behaviours elsewhere. To keep us on track in this book, I'll dig no more into these problems, reiterate the principle to 'not be a jerk', and conclude that overconfidence is dangerous, in every way it presents itself.

More positively, and something which is an utter joy to see people reach, is the 'just right' level of porridge'y impact confidence. It's that sweet point of balance where one is self-assured in their expertise but also open to learning, correction and humility, particularly learning from those with lived experience. It is by far the best level of expertise to hold, because none of us are ever the only expert in the room. We know stuff, we strive to learn stuff and we do our damnedest to deliver research with the most integrity possible, but as soon as you step into impact you immediately find yourself in a partnership with others. That might be with a named person, business, charity or other group entirely or someone *as yet unnamed*. Whether they're close by or still in silhouette, you're in an impact relationship, and recognising what they know, feel and need is key to meaningfully unlocking impact. And when it comes to thinking about the future audiences for impact, the great philosopher Michael Bublé had this pretty sewn up in the lyrics 'I just haven't met you yet'.

> 'At York University (Toronto, Canada) we established a knowledge mobilization unit (KMb York) in 2006. One of the primary functions is to support collaborations between non-academic organizations seeking to partner with academic researchers on a topic of shared interest. Once we have helped a non-academic organization develop a research question we need to find a researcher who might be interested in collaborating. Before we put a researcher in front of a potential research partner we check that the researcher appreciates that they 'don't know it all'. They need to recognize the complementary expertise of lived experience or community members. If they don't then we leave them to do their excellent, traditional academic research and we don't try to force them into a knowledge mobilization partnership.'

> Dr David Phipps, Assistant VP Research Strategy &
> Impact, York University (Toronto), and Director of
> Research Impact Canada

Thankfully in a wonderful antidote to the hot porridge jerks, there are far more decent, warm, kind and supportive people ready to help people hit that 'just right' balance. For all the jerks that might try to steamroller or block our efforts, there is an army of people pummelling the system with calls for integrity and fairness. Woefully the Goldilocks battle rages on in various ways, fuelled by personality and contexts, but where impact is concerned there's now many more allies ready to deploy collaborative support. Personally one of the most joyous things about writing this book has been reaching out to some of that army, and having them not only contribute, but to see them do it with such energy, faith and wisdom. And some heckling, but mainly the other things. If you don't have a just right porridge army, keep looking because they are worth their weight in gold.

SUMMARY

If you had been considering becoming a jerk I'd like to end this chapter with a suggestion to, well, not to. The simplest way to shift your thinking – and therefore your behaviour – is to remember that expertise is both multiple (rather than singular) and the domains of both academia and the wider world. Academic knowledge is fantastically valuable – embrace it, grow it, bathe in it if you want. It's just that for impact it's only part of the picture, and needs combining with the expertise of others to drive research into use. The service user who understands the difficulties of access. The teacher who understands a crowded curriculum. The industrial lead who knows the regulatory landscape. The artist who knows what challenges are finding galleries to exhibit in. And the school student who can remind you that however legitimate it is in music psychology, the term 'arousal' carries a fully different and hilarious meaning to teenagers[1].

My advice to you: sit up proudly, know what you know and recognise what you don't. Park qualifications at the door and listen to both what matters and what can (or can't) be done. Be confident in what you bring to the party, and make space to listen and learn from those who bring something else. Be humble, be open and bring biscuits. Call out jerks. As soon as you recognise that impact is dependent upon these relationships, you open the door to genuine change.

Just decide for yourself that you're not going to be a jerk to the people inside or outside of academia or basically any human being you meet. If you disagree with this principle, there's probably not a lot I or anyone else can do to help, and let's face it, impact is probably not your main problem.

WHAT CAN YOU DO?

Recognise and be proud of your expertise, welcome the expertise of others, engage openly and listen well. And don't let the conference coffee go cold.

---

[1] Not my best ever teaching decision

Ask yourself:

- Am I clear on what expertise I bring?

- Am I clear what expertise others bring?

- Am I under, over or just-righty confident in my expertise?

- Am I, or am I considering becoming a jerk? If no, continue. If yes, I look forward to hearing *it's not a question, more a comment* from you at a future event.

Principle 7

BE AN IMPACT LIGHTHOUSE

Bring your impact vibe to everything

You might have gathered by now that I'm a fan of an analogy. So let me explain this one.

Impact is – too commonly to be healthy – contained within discourses of assessment or funding. We have unitised impact into countable 'case studies' and there's still routinely enough of a last-minute rush to write impact plans in bids such as to underscore how 'extra' impact often still feels.

The problem is that impact itself doesn't just exist in these places. Impact needs the energies of researchers, research managers, policy-makers, communicators, the public ... we might contain the writing about it to these places, but the actions and the importance sit well beyond these points. And impact is weaved throughout so many of the processes of research – reviewing bids, recruitment and more – that locating our thinking to these places thoroughly overlooks the reality of how it works.

I've talked at length about the issues of unhealthy practices, and how crucial it is that we build healthy environments for impact. But there's another thing we can do, even when we're not in a position to change the institution or the infrastructure around us. We can build our own literacy and principles of practice, bring good understanding of impact to everything we do in academia, and help guide others who are struggling to find their way. We can bring to

bear our knowledge and understanding in a way that stops ourselves and others crashing on the impact rocks. We can be *impact
lighthouses*.

And yes, this principle is the antidote to being a jerk.

WHAT TO ILLUMINATE

It can be so easy for impact to be put under the domain of one
person, one department or even only fixable by external 'saviours'.
But we already have *so* many of the things we need to do impact
well. Academia is full of AMAZING people. People whose ideas
are beautifully creative and dumbfoundingly brilliant. Whose
ability to wrestle the chaos of complex projects into manageable
threads is awe inspiring. Whose kindness helps us through the
toughest of times and whose humour keeps us going when we've
got nothing left in the proverbial tank. Of *course* this isn't enough
whilst we wrestle with issues of equality, burnout and all else, but
that doesn't mean we shouldn't celebrate what we have, as well as
identifying what we need to change.

As soon as we refocus our lens to look at impact as part of the
research world, rather than just a thing contained in funding bids
or assessment, we can see the opportunities to shape healthier practice all round. And that means shining an impact-literate light to:

- Drive meaning-led approaches to impact.

- Resist Unicorn chasing.

- Build healthy environments.

- Identify or create opportunities for research to 'power up'
  change.

- Amplify efforts to collaborate, coproduce and engage.

- Build skills, knowledge and understanding.

- Address issues relating to privilege, power and wider ethics.

- Open the door for others.

WHERE AND WHEN TO SHINE THE LIGHT

Academia is a glorious mix of different roles and opportunities, meaning there are many ways and places you can shine a light. Here's a few ideas:

When planning impact: Plans at the outset of an impact journey set the tone and destination for what we go on to do. If you're in a position to support impact planning, shine a powerful light on the importance of engagement (and coproduction), establishing the impact need, designing realistic and achievable pathways towards change, and identifying impact goals which are meaningful rather than fantastical. By holding true to these values, you can help others' see through the impact mists and design something that could really make a difference. *Goal: help build real plans for change.*

'If we don't make good impact plans, how can we know where to head? And if we don't make plans with the people who're affected, how can we know we're doing the right thing?'
Dr David Phipps, Assistant VP Research Strategy & Impact, York University (Toronto), and Director of Research Impact Canada

When managing impact: Managing impact relates to guiding and overseeing an impact journey, commonly in relation to a plan that's already been made. This might be done by researchers, research managers or anyone else in a project management or oversight role. Impact lighthousing in research management can help balance the need for oversight with challenges borne from the inherent uncertainty of impact. Not only does that stop people from getting shirty that an impact plan made two years ago isn't working out as planned, but it can underpin long-term monitoring to see where the journey goes. *Goal: bring structure and flexibility to the oversight of impact.*

When creating impact strategies and implementation plans: Research strategies which mention impact are important, but aren't necessarily enough to create the conditions for impact to become reality. By shining your impact light on the strategy process, you can ensure plans avoid Unicorn chasing, recognise the skills and labour needed for both impact and evidence, and embed healthy practices across the organisation. *Goal: shape a strategy which not only amplifies impact, but brings everyone along on the ride.*

In supervision and mentoring: Especially for those in later career roles, there is a fantastic opportunity to help those newer to research to develop clear and healthy approaches to impact, cognisant of but not bound by formal agendas. Helping them to consolidate their thinking on the *WHAT, WHO, HOW* and *WHY* of impact *for their research,* recognising their contribution to the bigger picture, and doing this within the wider pressures of academia is pivotal if we're going to support individuals and establish a legacy of healthy practice. *Goal: help people find their place in impact.*

"As a PhD student, impact wasn't something I 'had' to do, but I felt driven to make sure that my research mattered. I wanted it to make a difference, and because my research area was sexual violence within higher education institutions, it was important to find ways in which my work contributed to the 'bigger picture' that is preventing violence against women. This happened for me through connecting with others in the sector seeking to make change too, and by having a supervisor who understood the importance of research to drive change. Without the investigation through research, true change would be much slower – and they reminded me of this when I felt that my work was too narrow, insignificant, or not enough to tackle such huge social problems. It was vital for me to

*have colleagues with more experience of research and its
translation to help me navigate these impact waters while
I also set off on developing my skills as a researcher."*
Dr Rebecca Brunk, Early Career Researcher, and previously PhD student at the University of Lincoln

When leading research and impact: If you are in a leadership position, then being an impact lighthouse is critical.
Unlike others in the system, you have more chance to create
healthy conditions, address resourcing and capacity, consider
skills development and set the overall vision for 'what matters'. Not only that, but depending on the nature of your role
you may also be able to look more holistically at the skills
and opportunities within your team and spot ways to sweep
research into teaching, consultancy or other practices. As a
leader, your guidance is paramount, and by understanding
how impact works you can more effectively and more collectively make it happen. *Goal: create the safe conditions for
driving impact.*

When acting as an impact champion: For those of
you working in direct support of impact, perhaps as an
impact lead, impact officer or erstwhile titled 'champion' of
some type, you're already a lighthouse. Your role exists to
be a credible guiding force through this thing called impact,
helping people avoid the rocks and get their research boat
to where it needs to be. By not only recognising the inherent
variety of impact paths, but also being able to help people
judge which way to go, you offer an extremely important and
timely contribution to meaningful social change. *Goal: be a
navigator, keep doing what you do.*

When supporting knowledge exchange: If you work
in the more business related or knowledge exchange end of

research, being a lighthouse means aligning the opportunities available, what matters (to both the institution and the external partner), with actual change (impact). It can be easy to focus on the process of connecting without anchoring plans in what can change or how the benefits can be measured. It can also be easy to see the fantastic opportunities through knowledge exchange (KE) without necessarily recognising how their value gets masked by the terror of trying to do more in a working day. By bringing impact literacy to your KE expertise, you can not only strengthen the effectiveness of KE for impact, but also help build strong two-way communications with academics to understand how to healthily optimise these opportunities. *Goal: Embed impact thinking into KE to supercharge both.*

"KE is brilliant for impact, but sometimes it can be easy to put the foot too heavily on the KE accelerator. If colleagues or research aren't ready to work with businesses, or if other pressures are too much, what looks to be a fantastic opportunity for KE colleagues can feel terrifying to researchers. To me it always feels like that famous Cheese Rolling race where a Double Gloucester is pushed down an English hill chased by a fearless crowd. The cheese gets to the bottom regardless, so what we need to do is reduce how injured people get as they chase it. Researchers, be clear with the KE office what you want, and KE colleagues make sure people feel comfortable saying if something doesn't feel right. People don't join academia to do a Knowledge Transfer Partnership – we need to understand how it fits into their career."
Helen Lau, Associate Director of Knowledge Exchange,
Coventry University

When reviewing impact plans or cases: Reviewing is commonplace in academia, be it for articles, funding bids, case studies or anything else we produce. Reviewing impact though can feel particularly taxing, especially if you don't feel you know what you're looking for. This is where your impact lighthouse can shine through: in impact plans,

pressing the need for researchers to not only be clear on
significance and goals, but also ensure engagement with non-
academics to determine clear routes to address 'what matters';
and in assessment, extending this to scrutinise claims, corroborating evidence and both reach and significance *in context*.
By shining your impact light you can imbue these processes
with fair and meaningful approaches which avoid unrealistic
goals and disfavour unreasonable BSEs. *Goal: ensure impact
is reviewed in relation to what matters in context.*

When writing case studies: We outlined what a
stronger narrative consists of in Chapter 1, but it
can still be hard to know quite what to write within page or
word limits. Your understanding of contribution, attribution,
evidence and directionally can add huge value to this process,
authoring or helping authors to commit the key aspects to
paper in the most legitimate, substantiated and impressive
way possible. For so many people, reducing years of research
and impact to a few pages can be both daunting and disillu-
sioning; your lighthousing can help them make a powerful yet
authentic story. *Goal: help bring structure, significance, reach
and authentic claims to the fore.*

In impact information management: Where impact
needs to be recorded, such as for assessment or for
funders, it is very typical to have some kind of system in place
(internally developed or purchased from a third party) to
manage the information. Nothing wrong with that, but the
problems start if your 'data entry' relies on people understanding what impact is. If colleagues are required to log
evidence of impact, how would they know what counts? And
if they are free to put everything in, how would they know
when to stop? Systems which rely on individuals to triage
what is or isn't impact also therefore require people to understand impact. Successful impact information management
needs impact literacy built in not only about the materials to

be added but in the way impact exists. By bringing your light-housing into the process, you can help ensure systems match the way impact works (for example, recognising changes may happen at any time, from the findings or the process, with singular or multiple pieces of evidence needed to tell the story), that guidance marries the realities of doing impact, and that the system can help support 'neat' reporting needed to monitor what impact is building. *Goal: help design what should be managed, and help people know what to provide.*

When collecting evidence: Evidence gathering may be done by academics, research officers or those employed for that precise purpose. For impact it doesn't matter who's collecting it, but if you are, shine your Murder She Wrote'esque light (Principle 4) to illuminate what evidence can be gathered, what it might tell us, how we might corroborate complicated claims and how a tapestry of evidence can be a viable method. *Goal: create a legitimate set of evidence that stands up to scrutiny.*

When developing a career: Ok so we're all doing this in *some* way, but it can be tricky for those not involved in impact to see how it features in a career. But by decoupling impact from the documentation we contain it in, we can shine our light on how we can contribute to its *purpose*. This could include for instance attending impact training, being involved in relevant professional networks, producing non-academic outputs designed to drive change, participating in knowledge exchange activities, integrating research into our own practice, generating impact from our own research or any other action that marks a commitment to making a difference in the world. Whilst I'll obviously advocate pursuing an impact career because impact is cool, for most people it will just be *part* of a fuller professional identity. *Goal: work out how you can connect with impact, in a way that shows development, effort and advancing skills.*

'*One of the key lessons that has emerged from the UK impact and KE agenda is that partnership working is valuable at all stages in the research process, from conception to dissemination. It is notable that our research training structures, particularly for early career researchers, are only just catching up with this reality. Much of the work of the impact support professional is seeking an appropriate balance between push factors – the dissemination of completed research, and pull factors – the social need, real or imagined, for different forms of research output.*'

Dr Chris Hewson, Faculty Research Impact Manager (Social Sciences), University of York

SUMMARY

Being an impact lighthouse is about illuminating how impact works and how it can be healthily integrated into our mixed academic lives. It is about standing true to fairness, meaningfulness and realistic expectations, and helping guide those with less experience or exposure to impact to do it 'well'. The opportunities to be a lighthouse are endless, but often not immediately obvious, so make the decision to do it whenever you can, and the opportunities will find you. I will delighted if anyone chooses to do this by dressing up in a stripy cape and a headlamp, whilst turning slowly on the spot.

WHAT CAN YOU DO?

Shine your light whenever you can, as brightly as you can, and in as many directions as you can. And look for others who can light your way too.

Ask yourself:

• Where in academia do I connect with or somehow touch on impact?

- Where and how could I be an impact lighthouse?
- What aspects of impact could I most strongly shine a light on?
- How can I boost the beam?

Principle 8

BE YOU

Within all this, be you. Unless you're a jerk in which case be someone else.

#liveyobestlife

Lorna Wilson

FINAL WORDS

Impact is the provable benefits of research in the real world. It is the most powerful term we have in for the contribution our research makes outside the academic walls, rather than the activities or any accolades within academia itself. Impact is so often driven by the demands of assessment and requirements in funding, but for so many people it's a personal motivation to make a difference that keeps us going. Whatever the impetus for impact in your world, or the rules and regulations you might need to adhere to, impact is and always will be about making a difference that matters.

Developing an impact literate mindset then is about locking into the sense that research can underpin or contribute to real world change (*WHAT*), via various paths (*HOW*), by engaging with different people (*WHO*), and in ways that are meaningful outside of the scholarly bubble (*WHY*). That might need baton passing over a long race, or you might already be near the finishing line. It doesn't matter; what matters is that you find and *power up* the links in the chain where your research contributes to making the world a better place.

Impact literacy is about starting from what matters and working backwards. If we are to do this sincerely, and in a way that reflects the opportunities and challenges of impact, we need to set some guiding principles. For me these are:

- *Chase meaning not Unicorns*, centring our thinking on what matters to those outside of academia, which absolutely, 100%, completely and utterly means connecting with people.

- *Work out what your research powers up*, recognising how different parts of our research can be made useful, for different reasons, and through the energies of different stakeholders.

- *Think directionally not linearly*, shifting from thinking about impact as a straight line and instead a series of directions linking 'problem' to 'change'.

- *Evidence? Think 'what would Jessica Fletcher do?'* Thinking like a detective to gather evidence and corroborate to prove impact as conclusively as possible.

- *Build healthy environments*, wrapping a supportive structure around the people and processes involved in impact.

- *Own your expertise but don't be a jerk.* Because, well just because.

- *Be an impact lighthouse*, guiding people, weaving healthy approaches into everything we do, and stopping people crashing on the impact rocks.

- *Be you.* Be authentic. Because there's enough fake bravado around.

The world can be a complex, changeable, varied, glorious, frustrating, exhilarating, overwhelming and unfathomable big round thing. The aim of this book has been to equip you with an understanding of impact and some guiding principles to help you do it in practice. It's also been to remind you that impact is fundamentally about working with people and that's really at the heart of things. Making a difference to someone's life or the world around us is one of the biggest privileges we can have in research, irrespective of formal agendas. The pressures and pace of academia can easily divert our attention from principles and idealism towards mechanistic behaviours of compliance, but this book is designed to help you hold true to those values even when life is busy. It's my hope that if you take nothing else from this book, you choose to anchor yourselves in meaningfulness and (continue to be) a fully decent human being.

It's entirely possible you think differently about impact, or think any criticisms I've raised in the book are unwarranted. That's entirely your call. Partly that may be because no single text can do justice to the complexity, nuance and multidimensional nature of impact, but it may also just be a difference of opinion which reflects the glorious tapestry that is the world of impact. But if you want to disagree and risk the wrath of Jessica Fletcher, that's on you.

So embrace the opportunity, accept the privilege and enjoy the journey. Be excited, be committed, be passionate, be annoyed, be productive, be unproductive, be an advocate, get chips, watch a play, get distracted by adverts about kitchen storage ideas (just me?), dance into the night and shout at the TV. Just don't get bogged down. Buoy yourself up and know you can make a difference. Find the need. Look for solutions. Bring your vibe. Be a lighthouse.

Impact matters because the world matters. And that'll do me.

Good luck x

FREQUENTLY ASKED QUESTIONS

What is impact?

The provable benefits of research in the real world. See *Chapter 1: What Is Research Impact?*

What is knowledge mobilisation?

The act – through whatever means are appropriate – for *actively* connecting research with society. Related terms (and often used interchangeably) include knowledge exchange, knowledge transfer, knowledge brokerage and engagement. Each has its own connotations, and different agencies (e.g., funders) may have specific requirements, but ultimately its about connecting with the world not just shouting at it.

Who decides what impact is?

Ultimately only those affected by research can say what the impact is, but funders, assessors and anyone else setting an agenda may add eligibility criteria or specify scope within a particular context. See *Chapter 3: Impact, Values and Power.*

What is impact evidence?

Information, data or other means to corroborate impact claims. Sometimes it's very clearcut, sometimes you need to assemble a case from a suite of different complementary pieces of evidence. Ultimately evidence is anything that stands up to scrutiny to show the impacts happened, and happened in the way you say. See *Principle 4: Evidence? Think – What would Jessica Fletcher do?*

Should all research have impact?

No. That's somewhere between lunacy and arrogant impossibility. Some research is very early in the 'translational chain' and thus unlikely to have an immediate application, and some is dependent on factors outside its control (e.g., funding available for the research to be used in education). My stance (others are of course available) is this: we have a duty to make work accessible and visible, and where it can have impact it should and we should drive it. Academia is a powerful engine and so we should use it to make the world better. But the system also needs to recognise impact may not be realistic in some circumstances. See the 'Disciplinary Differences' section in *Chapter 1: What Is Research Impact?* and *Principle 2: Work Out What Your Research Powers Up.*

What is better impact?

The short answer is 'whatever makes the most difference for those to which it matters'. The longer answer is 'who's asking?'. See also 'What Counts as Better' in *Chapter 1: What Is Research Impact?*

What if my work could have loads of different impacts? How do I choose?

This is often a fantastic position to be in; the issue is one of prioritisation. The decisions you make will necessarily be dependent on the context, but are typically related to judgements about opportunities, resources, need and impact goals. See 'What Counts as Better' in *Chapter 1: What Is Research Impact?* and 'Prioritising' in *Principle 2: Work Out What Your Research Powers Up.*

Does impact have to be in the same domain as the research?

Absolutely not. Impact can be in the same domain (e.g. social research having social impact), cross domain (e.g. arts research having environmental impact) or multi-domain (e.g. business research having social, economic and cultural benefits). There is precisely no set rule on what kind of impacts your research can or should have; it's all down to the topic and the context. Enjoy.

What do funders want?

All funders are different, and at the point of writing not all funders require impact as part of a bid. That notwithstanding, most funders will highlight their impact vision in a mission statement (check their website) and focus this down further in a particular programme or call. This might not be explicit, but use your impact spidey-sense to work out what they see as *benefits*, the types of award they are looking to make (e.g. exploration vs. intervention development) and what instrumental, conceptual or capacity impacts they might applaud. Ultimately, with funding ever more limited and competitive, funders want to know that monies awarded somehow convert – directly or indirectly – into benefits to society. See 'Impact in Funding' in *Chapter 1: What Is Research Impact?*

How do I review impact in funding bids?

Many of us review funding applications, be they internal awards or those for major funders. When we judge applications we typically fall back on our disciplinary expertise to know if it's good research uses strong methods, etc. Assessing impact at the planning stage needs a tweak to this, focusing not so much on the kudos of the methods, but rather how the plans *for implementation* reflect a clear need, with realistic pathways and suitable engagement towards demonstrable impact. Impact is inherently uncertain at the start, so judgement needs to balance the potential benefits with any barriers or facilitators along the way. Impact reviewing is primarily an assessment of achievability, practicability and meaningfulness. See 'What Counts as Better' in *Chapter 1: What Is Research Impact?* and 'Impact in Reviewing' in *Principle 7: Be an Impact Lighthouse.*

What does assessment want?

Impact assessment is the process of marking 'what has happened'. Rules and requirements will vary by who's assessing, for what purpose and in what way, and so it's crucial to check the rules you need to comply with. Typically assessments have criteria around eligibility (time frame, institution or other such limiters), and specific requirements for submission (format, structure and evidence). See 'Impact in Assessment' in *Chapter 1: What Is Research Impact?*

Is impact-for-assessment the same as impact?

No. Firstly impact existed before formal assessment agendas (I hope you were sitting down for that mind exploder) and assessment is a selective process. Impact is the provable benefits of research in the real world; impact assessment typically looks at a selection of examples which align with eligibility requirements around (e.g.) dates, staff contracts, judgements of research quality within that field.

How do I write a great case study?

Good impact narratives tend to share certain characteristics (See 'What Counts as Better' in *Chapter 1: What Is Research Impact?*), but it will always be necessary to check the guidance for the specific scheme you're submitting to. In general, stronger case studies tell a clear story of research contributing to significant impact(s), show more strategic, 'high level' or deeper changes, and have clear and verifiable qualitative or quantitative evidence.

How do I review case studies?

Carefully. In assessment cycles, there are typically a series of internal review points of potential case studies as they grow from early to more mature states. In the early stages, reviews are best focused on identifying potential impacts, how they can be nurtured, how likely plans are to come to fruition, and what could be monitored to provide evidence. As the cycle runs on, this will naturally shift to considering the fuller narrative to be told, the scale of impact and evidence, how effects can be strengthened further, and any gaps to be filled. Towards the end, reviewing shifts far more to the strength of the story, the connection of this to the evidence and the overall 'believability' of what's being claimed. Throughout, remember to focus both on what impact the case is conveying and how they are doing it. You need good content AND a plausible, attributable, connected story. See also 'What Counts as Better' in *Chapter 1: What Is Research Impact?*

What if I do fundamental research?

It's important we protect the value of exploratory work, without suggesting that such research can't or shouldn't have impact. Where it can it should, and that may need more collaborative working or longer term vision. See the 'Disciplinary Differences' section in *Chapter 1: What Is Research Impact?*

What if I do practice-based research?

Then you're already super close to impact. Typically the issue for practice-based researchers is delineating the research from the effect, because the interaction with society is continual, iterative and reformulating. The impact doesn't just come after the research, but can come through the process itself through all those lovely interactions. See 'Disciplinary Differences' section in *Chapter 1: What Is Research Impact?*

What if my impact is mainly through public engagement?

Brilliant, public engagement is a powerful part of the impact process. Public engagement is about sharing and democratising knowledge, and bringing insights back into academia to inform what we do. But it can also be challenging to prove impact from public engagement as the audiences can range from those captive at a talk through to people who might glance at things on social media. Think about how active your audience is (see 'WHO' section in *Principle 3: Work Out You're Your Research Powers Up*), and consider what evidence you might be able to use to show change (see *Principle 4: Evidence? Think – What Would Jessica Fletcher Do?*).

How do I build a healthy institutional culture?

See *Chapter 5: Create a Healthy Space* for fuller coverage of this, but in essence you need to address institutional commitment, connectivity between teams, coproduction with society, competencies across the organisation and clarity on what's expected. Within this, you also need to cement a clear understanding of *WHY, HOW, WHAT* and *WHO* and unpack what resistance is actually reflective of. And bring biscuits. The good ones. You know, individually wrapped.

Should I prioritise what I can measure?

Only if you're a tape measure.

Is open access and open science enough for impact?

Openness will always help, but it it's rarely enough. Available doesn't mean accessible, and materials need to be understandable, usable and focused to the audiences who need them. See the 'Who' and 'How' sections in *Principle 2: Work Out What Your Research Powers Up*.

Do journal impact factors demonstrate impact?

No. Nein. Non. Ne. Nee. Nope. Impact factors are an increasingly frowned upon bibliometric for showing the attention for a particular journal and as they don't reflect real world change can't be marker of impact. Seriously. See 'Things That Sound Like Research Impact But Aren't' in *Chapter 1: What Is Research Impact?*

What if I hate working with people?

Assuming the problem is you and not them, then you're going to need to have a word with yourself. See *Principle 6: Own Your Expertise But Don't Be a Jerk*. If you have imposter syndrome, recognise the value you bring alongside colleagues and with those outside of academia. If you think it's unreasonable you should have to, or that people are annoying and just get in the way, dial down your inner Sheldon Cooper[1] and just listen. Draw on the guidance and expertise of your colleagues working in public engagement, communications and business development. Do some media training

---

[1] The glorious theoretical physicist from The Big Bang Theory TV show, if you're not familiar with his work.

and get involved in outreach and knowledge mobilisation. Or like Sheldon, and this is probably a last resort, project your face onto a tablet strapped to a robot and get that to interact with people instead.[2]

How can I engage in impact as a student?

Other than any specific eligibility criteria in funding or assessment, there is nothing stopping you. Speak to staff, connect with people outside the institution and don't be scared to think about how you can start the chain with your own research.

How do I demonstrate impact in career progression?

The way impact features (or doesn't feature) in the appraisal process is completely varied, and typically reflective of how substantially impact is formally expected. You'll need to establish what is included in your progression criteria, but starting points would be to use the skills list (see the competencies section in *Principle 5: Create a Healthy Space*) to articulate what you contribute, and log any activities in support of impact development (see *Principle 7: Be an Impact Lighthouse*).

Should I align myself with a specific political agenda to drive impact?

Oh that's rarely a good plan.

What's the main thing I should know about impact?

It matters, because the world matters and we have a chance to make it better.

---

[2]If you've not watched The Big Bang Theory, this will appear to be hallucinatory. But feel free to google 'The Cruciferous Vegetable Amplification' Series 4, Episode 2.

INDEX

Note: Page numbers followed by "*n*" indicate notes.

Academic citations, 13, 31–33, 79, 161
Academic impact, 30
Academic influence, 22
Assessment, 12, 18–19, 23, 25, 29, 31, 34–35, 43, 46–48, 63, 68, 70, 80, 107–108, 140, 150–151, 155, 173–174, 179, 186
  cyclical, 21
  formal, 141, 143
  funding and research, 18
  impact, 20–22, 70, 86
  REF, 21
  research, 38, 122
Attribution, 37–38, 46, 179

Beneficiaries, 19, 126, 154
Better impact, counts as, 45–47
Bibliometrics, 13, 31–32
Big Shiny Endpoints (BSE), 63–65

Capacity building impacts, 15, 124
Career, impact in, 180
Case studies, 21–22, 47, 51, 61, 63–65, 70, 78, 80, 86, 90, 151, 168, 173, 178–179
Citation God Complex, 32
Civic Universities, 24*n*12
Clarity, 26, 30, 57, 146–147, 151–152, 154, 161
Commissioned research, 44
Commitment, 52, 75, 78, 140, 144, 151, 153, 155–156, 161, 180

Competencies, 152–153, 160–161
Conceptual impacts, 16, 124
Connectivity, 151, 153
Contribution, 9–10, 14, 20, 24, 30–31, 37–38, 46, 49, 71, 74–75, 77, 90, 124, 134, 176–177, 179, 186
Coproduction, 10–11, 29, 55, 75, 95, 105, 152–153, 159, 161, 175

Dimensions of impact, 34–40 (*see also* Reach, Significance, Contribution, Attribution, Distance, Time, Linearity, Dependencies)
Directions of impact, 112
  from 'problem' to 'better', 112–113
  baseline, 114–116
  impact goal, 116–118
Distance, 29, 38–39
Domains of impact, 15, 40
Dual funding system, 18, 52

Embedding impact culture, 154–162
Equality, Diversity and Inclusion (EDI), 68
Evidence, 11, 17, 121–136, 180
  collecting evidence, 180
  corroborating impact, 135
  hard proof, 124–129
  impact, 123
  logical proof in uncertainty, 127–128

proxy measures, 126–127
softer proof, 125
using events as evidence
    waypoints, 128–129
what counts as evidence of
    impact, 129–134

Failure, 65–67, 70, 80, 139, 167
Five Cs of Institutional Health
    151–154 (*see also*
    Commitment,
    Connectivity,
    Competencies,
    Coproduction, Clarity)
Frascati definition of research, 12
Fundamental or 'discovery'
    research, 41
Funding, impact in, 18–20

Goldilocks Paradigm, 166

h-indices, 31
Healthy institutional culture, 149
Hidden REF programme, 74

Impact champion, 146, 177
Impact culture, 66
    embedding, 154–162
Impact lighthouse, 173–182
Impact literacy, 1–2, 4, 20, 51, 53,
    60, 149, 186
    core elements of impact literacy,
    55
    evolution of model, 53–56
    HOW, 54, 144–145
    institutional impact literacy,
    142–145
    instrumental impacts, 15, 124
    levels of literacy, 58
    mindset, 1–2
    risks of taking non-literate
        approach, 56–58
    WHAT, 55, 144
    WHO, 56, 145–146
    WHY, 55, 143–144
Impact terminology, 25–34, 103
Imposter syndrome, 167
Information management, 179–180

Institutional health, 149–151
    (*see also* Clarity,
    Coproduction,
    Commitment,
    Competencies and
    Connectivity)
    5Cs of, 151–154

Jessica Fletcher, 121–135
Journal impact factor (JIF), 30–31

Knowledge exchange (KE), 4,
    25–26, 28–29, 111,
    155, 158, 177–178, 180
Knowledge mobilisation (KMb),
    19, 26, 28, 60, 68,
    129, 160

Leading impact, 177
Linearity, 39–40, 111–112, 118

Managing impact, 175
Mentoring, 176
Missions, impact in, 22–24

Participatory or engaged research,
    42
Personal motivation for impact,
    24–25
Philosophical research, 42
Planning impact, 175
Powering up, 94, 96, 102,
    107–108, 128
    what can be mobilised, 94–96
Pressures on people and
    institutions, 70–75
Prioritising, 87, 107–108, 122
Public engagement, 2, 28, 104,
    106, 125, 129

Reach, 35–37
    horizontal reach, 35–36
    reach as depth, 36
    reach over time, 36
    vertical reach, 36
Relationships, 159
    breaking up, 100–102
    partnering, 98–100

Research in contested, sensitive, taboo or secret areas, 43–44

Research to curate, preser ve or order knowledge, 45

Research to develop useful 'thing', 42–43

Resistance, 140–142

Reviewing impact plans or cases, 178–179

Significance, 34–36, 46–47, 107–108, 130, 179

Stakeholder energies, 97

Strip clubs, 65–66, 70

Supervision, 176

Third Mission, 17

Time, 22, 25–26, 38–39, 48, 52, 60, 67, 75, 79, 89, 107–108, 141–142, 159, 167, 180

Types of impact, 14–16

Unhealthy cultures, 151

Unicorns, 86–87, 91

harnessing unicorn energy, 89–90

United Nation's Sustainable Development Goals (United Nation's SDGs), 23–25, 31, 48